"You have to come up with an idea, work it through, and only then figure out how to test it experimentally. In that sense, science is an art."

Albert Einstein, April 1950

GENES AND HUMAN NATURE

* * * * * * * * * *

... *FROM ATOMS TO "GOOD & EVIL"*

by

Dr. Hussein A. AMIN,
Professor of Surgery

Bloomington, IN  Milton Keynes, UK

authorHOUSE

AuthorHouse™
1663 Liberty Drive, Suite 200
Bloomington, IN 47403
www.authorhouse.com
Phone: 1-800-839-8640

AuthorHouse™ UK Ltd.
500 Avebury Boulevard
Central Milton Keynes, MK9 2BE
www.authorhouse.co.uk
Phone: 08001974150

First published by AuthorHouse 7/5/2007

ISBN: 978-1-4259-2651-9 (sc)

Library of Congress Control Number: 2007905011

Printed in the United States of America
Bloomington, Indiana

This book is printed on acid-free paper.

Dr. Hussein. A. AMIN, M.D., FACS.
Professor of Urological Surgery
44, NEHRU Street, Heliopolis, Cairo, Egypt.
Phones + Fax. (+ + 20 -2) 452 1075
                (+ + 20 -2) 258 1265

First paperback edition 2005.

Editorial revision by Bethany G. Ming, ESQ.,
Attorney and former teacher of Sociology at Columbia College.

Dedication

*   *   *   *   *

To my grandchildren

*   *   *   *   *   *   *

Today's world is changing in
an accelerating way, almost
by the day. I believe that in
three decades time, none of
my peers would recognize it
at all…!

* * * *

Preface

* * * * *

Why I wrote this book? I was an avid reader since my teens, particularly about human nature and science. It was a fascination to keep pace with the ever-expanding frontiers of science. All the time, I noticed a thread of continuity that bound three branches of science: The physical structure of the universe, the physical structure of life on planet Earth, and the nature of mankind. In the last two decades we had an accelerating explosion of our knowledge of DNA and genes. Suddenly, I realized that it all made sense, in a way which I passionately like to share.

It took me four years of careful research, revision, and re-writing. The aim was to present such heavy scientific material in simple words and pictures that are comprehensible to the lay reader. I feel proud of what I've achieved. Throughout the book, the discourse is science, science that can be tested in a laboratory, and seen under a microscope ... No Philosophy... No Metaphysics!

> > > > >

Mankind is defined not as the animals that have souls, but as the animals that can *INVENT* ideas, and then *TALK* about them, (inventiveness and language). Signs of these two momentous abilities are documented to have *SUDDENLY* started only about 30,000 years ago. Speculations about the cause of that fateful incident, ranged from Supernatural sudden creation, to Insemination by space aliens, down to Haphazard *blind* evolution. But the G-SAT Theory, detailed in this book, offers a *SCIENTIFIC* frame, for a master-planned, _purely physical process_, that initiated Mankind.

I have chosen the above sentence to be printed on the back cover. It summarizes one of the four most important messages delivered by this book.

The second message: is Genetic-Religiosity. Our brains are hard-wired in a way that allows a meditating mature human to feel God within himself, rather than as an outside abstract thought. Meditation *is* an important item in the rituals of *all* religions.

Mankind has a genetic language syntax common to every human. It becomes activated at two years of age. One's specific language is something that is acquired, depending on the society he is born into. Similarly, a *genetic* religious instinct is common to all Mankind. One's specific religion is simply *acquired*, according to the society he is born into. When this idea sinks into someone's heart, then inter-religious feuds look irrelevant indeed, as would inter-languages-disputes.

Modern science is sowing the seeds of atheism everywhere. The knowledge that "Faith" is somehow hardwired into the human genome gives Mankind an option to avoid the bleak spiritual vacuum of atheism. It is such a noble, (and inter-religious peace-making), notion that knowledgeable and conscientious people should strive to spread; and this I'll try my best to do.

The third message: is about the universality of the genes, in all forms of life, bacterial, plant, animal, and human. This has become a well known fact, and is taken as evidence of ancestral relationship of all forms of life. If this ancestry was achieved through blind chance evolution, it would have needed several billion times the whole age of the universe. Recent genetic science suggests a different route: a *pre-determined-encoded-route* that could act within the recognized age of the universe. This is exactly comparable to the remarkable degree of scientifically evident directionality, or even purposefulness, in the evolution of the Universe.

DNA is *NOT* synonymous with '*GENES*'. DNA is the ink and paper on which the encoded messages of the genes get recorded. Every year, we come closer to deciphering how these messages work. How they get switched on and off and how they get suppressed or activated? With this rapidly progressing scientific discipline, we are getting closer, by the day, to scientifically proving my pre-determined-encoded theory.

The fourth message: is about 'Genetic Human Nature'. There are a few dozen human traits that are proven to be genetic. Culture, or nurture, acts like the volume and tone controls of a factory-made television set. They can augment, or tone-down, any genetic trait. In chapter (10) I have bared the human self, not down to the skin, but down to the naked genes.

I argue that contrary to common belief, The *Devil* is not anywhere, he is inside everybody. Humans have a difficult time believing that, so they gave him a name, a symbol, and legends, and then started to train curses on his head, as if by doing that they wash themselves clean. *The same gene system that predisposes Man to feel God also includes the wiring that predisposes him to the Devil.* By identifying his real self, knowing the weaknesses, and the strengths, Mankind becomes better prepared to deal with each other.

* * * * * * *

TARGET AUDIENCE:

The book is intended for men and women of all cultures and religions, in a popular way. I made a point of simplifying the language of scientific terms to suit someone, in an armchair with a shade lamp, enjoying a quiet evening with an interesting book.

REVIEW OF COMPETITIVE WORKS, AND HOW THIS BOOK IS DIFFERENT:

[A] Readers are fascinated by books which talk about how Mankind came to be. Like the idea of Space-Aliens inseminating our civilization 30,000 years ago, (Examples are: *Chariots of the Gods, God was an Astronaut, the Mars mystery,* etc.] This book is presenting scientific facts rather than layman's speculation, and will have an instantaneous appeal.

[B] Laypeople are hooked on the hotly publicized **genome** and its implications. The word 'Genetic' in the heading of any article or book is a sure attraction. But unfortunately none of the existing literature addresses our hypothetical reader, in the armchair, who may have a very limited background of technical knowledge.

[C] During the last two centuries, modern science has cast a very long shadow over all traditional religions. So much so that atheistic feelings have become almost synonymous with scientific knowledge. The shadow of atheism is further accentuated by the dogmatic-just-so -ism of most religions, by the conceptual contradictions that exist between some of them, as well as the fanaticism of their clergy. (The expression 'Just-So' means dogmatic statements for which no discussion is tolerated.) All this sharply contrasts with scientific thinking. On page 47 of Newsweek, July 27, 1998, Sharon Begley literally says that: *"Most scientists still park their faith, if they have it, at the laboratory door..!"*

While seemingly revolutionary, the message conveyed by this book is simple: scientists will not shyly hesitate to take their faith through the laboratory door anymore...!!

[D] Books talking about spirituality and the Divine are very popular. [Examples are: *Is God a virus? God, the Evidence, the Fifth Miracle, Why God Won't Go Away? Phantoms in the Brain, Does God Play Dice?* etc.]

Clergy, in all religions, who write defending religion, often fall into an interesting trap: their thinking marvels at the greatness of the universe and the complexity of our bodies. All this needs a Creator, they then say. Following that, with no exception, each starts to talk preferentially about his particular religious creed. Let us remember that all Christianity sects claim no more than one third of Mankind, Islam no more than one fifth, and Oriental religions even less. For devout people, such line of thinking is superfluous, for doubters and atheists it does not change their mind. Moreover, for the *majority* of six billion humans who were born to parents of *other* creeds, the ears, eyes, and hearts almost always get virtually sealed, starting from page 20, or even before. Our book, for the first time, talks about these matters not in the way of Scriptures, but in the language of science. It guarantees to keep the receptiveness of the reader aglow, from the first to the last page.

The *timing* of the book: It couldn't be better. The plague of falsely labeled inter-religious wars is spreading. The false synonymy of modern science and atheism is creating a huge spiritual vacuum. Its ominous results are manifest in every newspaper and magazine.

[E] **How do our brains work?** and the argument between whether brain function is physical, or the existence of a 'Soul' is an eternal fascination. [Examples are: *How Brains Think, the Mind-Body Problem, the Astonishing Hypothesis, Descartes' Error, the Emperor's Mind,* etc.]

The way this topic is handled in this book is quite different and simple. In a few pages, the reader will easily be able to discuss them with anyone who has waded through thousands of pages in other books.

[F] **The great debate between Evolutionists and Creationists.** It is raging everywhere and especially so in the USA, where it is hotly politicized. It is attracting readers everywhere. Look at the numbers of books sold every year dealing with this infamous fight. Both sides present nothing but equally unscientific *Just-So* arguments. Their victim is the ordinary reader, who gets terribly bewildered by the quasi-scientific jargon presented by both sides. Their real goal is not scientific truth, but billions of dollars in federal and private research funds. There is a special chapter in the book deciphering this debate down to its roots. Not only that, but also giving the unbiased scientific alternative to the dogmatism of both sides.

* * * * * * * * *

This book, [GENES and Human Nature] claims, on scientific grounds, that the evolution of life on Earth didn't proceed on haphazard-chance routes; but rather on a *pre-determined-genetically-encoded-line.*

This will be the first book to initiate a very important genetic EUREKA!!

The implications are great. The ripple it will start is expected soon to rise into a big wave. It won't stay for long as a theoretical idea, because, with the rapid advance of genetic science, the proof is both scientifically and practically feasible. It could be around the corner in less than a decade.

Hussein A. Amin,
Cairo, 2005

Editorial Comment

It has been my pleasure to work the past several months with Dr. Hussein Amin in preparation of this book. This book has been years in the research and writing by Dr. Amin, and is his first book written for the American and worldwide English reading audience. When I became involved in this effort, the book had already been completely written. It was my task to smooth out the language and structural issues that presented themselves as a result of being written by a non-native English speaker and writer. To his credit, Dr Amin, who has published several non-related works in Arabic and English for an Egyptian audience, wrote a well organized and clearly presented manuscript.

Throughout our work together, Dr. Amin stressed his desire that the book, although scientific in nature, be readable and understandable by the common person. This goal is accomplished by Dr. Amin's ability to explain science at a simple and popular level. It is further clear from his style of writing that Dr. Amin has a passion for this material and is eager to share it with the reader. Whether you are familiar with some of the concepts presented or are reading about them for the first time, this book collects and summarizes old and new ideas in a straightforward way.

The book is organized into eleven chapters whose contents build upon each other. Numerous subheadings and figures complement the text by providing easily readable segments and interesting illustrations. Index to figures, subheadings, referenced works and general index are provided.

Mrs. Bethany G. Ming
December, 2004

About the Author

* **Dr. Hussein A. Amin,**
  Professor of Urological Surgery.

* Graduation, Cairo University **1953**
Doctorate of Surgery, Cairo University **1960**
Fellowship of the American College of Surgeons **1966**
* In 1957 he founded and headed the Urology specialty in the hospitals of Kuwait, for 22 years. During his practice in Kuwait, (and hospital visits around the World), he published several surgical articles and books, on both local and international levels.
* Repatriated in Egypt in 1979 as Professor of Urology, and also shared in the foundation of the first large, (200 beds), multi-specialty private hospital in Cairo, inaugurated in 1981.
* Married with two sons and five grandchildren.
* An avid reader, for over half a century, particularly in the two challenging fields of: Modern Science, and Human Nature.

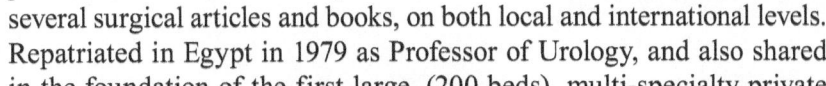

In 1991 he divided his day into two parts: in the mornings he is a very busy surgeon; in the evenings he enjoys, very much, reading and writing books and articles. He has seven successful publications...so far, and an eighth under publication:

1) *Money and Human Nature*, A panoramic book, in Arabic **1993**.
2) *Renal Stones, history, and prevention*, A book in Arabic, addressing the common man rather than the doctor. **1994**.
3) *History of private hospitals, and the future of medical Services in Egypt* **1995**.
4) *A new perspective of the 20th Century history*, An Arabic 74-week series in the Weekly Magazine *Akher- Saa* **1995 -1996**.
5) *Science versus Faith*, articles in *Shell* Magazine, Cairo **1995 -1996**.
6) *Feminism, the False Freedom!* A book in Arabic, as well as an article in English in a scientific magazine. **1999**.
7) *FEMINISM, a second look AFTER 50 YEARS*, a book in English, **2002**.
8) *GENES and Human Nature, from Atoms to "Good & Evil"* In English, under publication ... **2005**.

N.B. The author's name is proudly shared by two other writers, (and friends):
(1) Dr. Hussein Y. Amin, Professor of Information Science, American University, Cairo.
(2) Mr. Hussein A. Amin, Egypt's ex-Ambassador to Algeria, and son of the famous writer, and Minister of Education in the 1940s, Dr. Ahmed Amin.

CONTENTS

Index of Headings   xxi
Index of Figures   xxvii
Introduction   xxix

Chapter (1)   What is meant by GENES?   3

Chapter (2)   Who is Darwin,
                what exactly is the "Theory of Evolution"?   25

Chapter (3)   The history of life on Mother Earth   35

Chapter (4)   How we got COOKED, up there, in the stars?
                The Universe, the kitchen-work in the stars,
                Quantum & uncertainty principles   49

Chapter (5)   Why "GOD" won't simply go?   77

Chapter (6)   Co-Evolution & Adaptation
                The statistical probability of "chance"
                mutations, as the main force in Evolution   97

Chapter (7)   The Gene-Suppression-Activation Theory,
                (GSAT), as the main force in "Evolution".   125

Chapter (8)   Noah's Flood   149

Chapter (9)   Who is MANKIND?
                The 30,000 years landmark!!   157

Chapter (10)   Genetic Human Nature   175

                Human Nature and World Peace
                The DEVIL and Human Nature …!!   175

Chapter (11)   BODY, MIND, and SOUL …!!
                The personalized Human Brain …!!
                How is it made, and what does it mean?   223

References   247
Index   253

INDEX OF HEADINGS

* * * * * * * * *

PAGE

3-23   **CHAPTER 1**
          What is meant by 'GENES'?
5   Artificial selection, (breeding), Natural Selection, cells, Chromosomes, Mitochondria, Somatic cells, mitosis, Germ-line cells, Meiosis, DNA, Watson, Crick, Wilkins
7   The substance of the DNA chain.
8   How is heredity performed?
10   How does the DNA chain govern cellular function?
11   How does the Gene function?
12   What is RNA?
13   Language of the Genes, Genes as | different from DNA, Gene-suppression
15   Genomics
16   Study of the ageing process
17   Prediction of Gene-related diseases, Genetic finger-printing, Somatic genetic therapy
18   Pharmaco-genetics
19   Genetic engineering, Germ-line therapy
20   Stem-cell research, Free-Will versus genetic determinism, No, it is not all in the genes!
22   Genetic determinism versus religious thought

25-33   **CHAPTER 2**
          Who is Darwin? and what exactly is the theory of evolution?
27   Biography, the Beagle
28   Origin of species, Survival of the fittest
29   Galapagos finches
30   The false definition of "SPECIES", Micro-evolution, and its mechanisms, Macro-evolution
31   Peter and Rosemary Grant research, Alleged "mutations", The Manchester moth
32   Bacterial-Antibiotic resistance, Insecticide resistance
33   Haphazard random-chance mutations

35-46	**CHAPTER 3**
	The History of Lifeon Mother Earth
36	How do we investigate life on earth?
38	Story of the Earth
38	Tectonic Plates, Continental drift
39	Story of Life on Earth, Blue-green-algae
40	Panspermia, the Cambrian explosion of multicellular life
41	The dynasty of single-celled life, Selfish DNA
42	Viruses, The three blankets and balances of life on Earth
44	Earthworms
46	Summary of the recorded history of Mankind
49-74	**CHAPTER 4**
	How We Got Cooked Up, In The Stars
51	What is the universe, and how was the Earth formed?
52	A simple sketch of the universe
53	Einstein's Theory
54	How do we investigate the universe?
56	Our present knowledge of the universe
57	The dark matter and the dark energy
58	Where did we come from?
59	The kitchen-work in the stars!!
60	The source of energy in a star, The death of a star
62	Pulsars
63	The formation of heavy elements
64	Our solar system
65	Black holes
66	Chaos & quantum uncertainty principle
67	What if??!!
68	The anthropomorphic universe
70	The steady-state theory, Universe younger than its stars!
71	Earlier collapsing universe, multiple universes
72	Theory of the tailor-in-reverse
74	A conclusion comment for Chapter (4)
77-94	**CHAPTER 5**
	Why "*God*" Won't Simply Go?
79	Definition of 'Faith'
80	Common characteristics shared by all religions
80	The summary of today's main religions
83	Atheism as a Creation-Story

84   Religious fraud, Shroud of Turin, Donation
      of Constantine, Scientific fraud
85   The state of "Religion" in the 20th century
86   Resurgence of "Fundamentalism", in ALL   religions
87   The Evolutionists-Versus-Creationists debate
90   Scientific rules of the debate
93   The Principle of   DEISM, The "Listening" GOD
94   Religiosity,   as a "CODE" in the human genome

97-122   **CHAPTER 6**
          Co-Evolution, And Adaptation
99   Chlorophyll and hemoglobin
101   The start of sex
102   Echo-location, Technicalities of echo-location
104   The flying rats
105   Genetically-imprinted inter-species co-ordination
106   The funny story of Hydatid disease
108   Carnivores versus herbivores
109   Malaria
111   Bilharziasis
112   The European song-bird
113   Examples of symbiosis, Acacia trees and ants
114   Ruminating animals, symbiotic
      pollination, the hammer-orchid
116   The Rafflesia flower, A word of insight
      Symbiotic transport, the mistletoe
118   Camouflage and Mimicry
120   Discussion of alleged co-evolution & adaptation
121   The funniest way ever to end a book..!!
      Feasibility of chance mutations

125-147   **CHAPTER 7**
          G - SAT The Gene-Suppression-Activation-Theory
127   The "allegedly" junk-DNA
128   The temporarily suppressed butterfly, The
      alternately suppressed social insects
129   The eternally suppressed chimpanzee
132   The Christmas Island crabs
134   The scenario of life under this theory
136   Oxygen level as one factor, radioactive potassium, Ozone
137   The Gene-Suppression Theory and Mankind!

138   Human embryological evidence

141   Controversial questions answered by the G-SAT Theory: Cambrian explosion, Punctuation and stasis, Intermediate forms, Domestic animals, The look-alikes, Sets of genes, All-or-none-rule, Man's original sin, Time factor in chance-evolution

144   The Principle of DEISM, The dynasty of atoms

146   Post-script to Chapter (7), Excerpts from J.D.Watson's book "DNA, the secret of life", Excerpt from *"Scientific American"* magazine Oct. 2004.

149-155   **CHAPTER 8**
           Noah's Flood

151   The scale of Genesis, Creation forms

152   Noah's Flood, The many legends of the Flood

157-173   **CHAPTER 9**
           Mankind

159   Exclusive definition of Mankind

160   Language syntax, language instinct, archaic paintings and carvings

161   Creole languages

162   Written language

163   Inventiveness

164   Brain size, a false measure, Fossils of buried intelligence

165   Clothes

166   Definition of Mankind, The hominid line

168   The human mind and soul

169   Forces of Nature beyond our comprehension, Migrationsthe Monarch butterfly, The smell of salt

170   Was God an astronaut?

171   Book-selling fraud

173   The Raellian religious cult

175-220   **CHAPTER 10**
           Genetic Human Nature

177   The basics of human nature, the three essentials! Universality of human nature

179   he list of universal traits of human nature

180   Consciousness, Dreams of immortality, Selfishness

181   Narcissism, Envy

182   Tool usage and long-term planning

183   Private ownership

184   Wanting more, Vanity, Nakedness

185   Body display

186   Sex in humans, are we different? Sex versus love

188   Man's Achilles heel in the subject of sex, Sexual jealousy

189   Polygamy, The wedding celebration

190   Incest avoidance, Feminism

192   The concept of work & Money

193   Charity versus jobs

194   Grameen Banks, Menial jobs, Pride in work

195   Pride in emotions and love

196   The concept of Government, and lust for power!

197   Dictatorships

199   The concept of Free Will...and religious
feelings, Morality, Altruism

200   Mass psychology, When Mankind changes to 'Dinosaurs'!

201   William Golding, Bertrand Russell, atrocities during wars

202   Revenge

203   Forgetfulness and forgiveness, Aggression

204   Conquests as false historical honors,
false national pride, Homicide

205   Mankind, the social animal, Feeling needed

206   Herd approval, Tribalism, Feelings of tribal superiority

207   Sibling rivalry, Nostalgia

208   Ethnicity, Teenage rebellion, Globalization

209   Gossiping and voyeurism

210   Facial expression, blushing, tears

211   Boredom, Sympathy for the helpless

212   Natural aversion to snakes, Beauty, Music and Art

213   Natural Beauty, The feminine figure

214   Rhyme, poetry, music, dance

215   Human Nature and World Peace, The
"survival value" of human nature

216   The Devil and human Nature

217   The Devil in Evolution Theory

218   Dr. Jekyll and Mr. Hyde, and novel writing

223-245   **CHAPTER 11**
Body, Mind and Soul

225   Definition of Mankind, again!

227   Neurons, Interconnections, Sleep and dreams, Blood-brain
        barrier, Personalized brain, Happiness in childhood, and later
228   The effect of drugs
229   The brain's reward system, Endorphins,
        Dopamine, Nitric Oxide
230   The effect of injury, the effect of anoxia
231   The physical concept of consciousness,
        Methods of investigation
232   Methods of treatment, Fight-or-flight reaction, Chronic stress
233   The religious animal
234   Empathy
235   Morality in atheism, Freemasonry
236   Quackery
236   Consilience
237   Incest-avoidance and its significance, The Westermarck-effect
239   The Weird by-product!
240   After-life courts in Ancient civilizations, The dualism belief
242   Recent Scientific research into brain activity,
        closely related to the subject of Genetic-
        Religiosity, Ramachandran , Newberg

> > > > >

INDEX OF FIGURES

* * * * * * * * * *

FIG.   PAGE

(1)   1   /   The Old-World, a space-ship night view
(2)   9   /   Animal cells, chromosomes
(3)   11   /   Diagram of DNA and RNA
(4)   25   /   Charles Darwin
(5)   36   /   Mother Earth
(6)   45   /   The balances of life on Earth
(7)   48   /   The Universe
(8)   96   /   Two lovely migrating birds
(9)   110   /   Hydatid disease
(10)   112   /   The European song-bird
(11)   118   /   The hammer-orchid, and Rafflesia flowers
(12)   119   /   Camouflage & mimicry, 4 pictures
(13)   131   /   The suppressed butterfly, the suppressed chimp
(14)   133   /   Christmas Island crabs
(15)   140   /   Human embryological evidence
(16)   156   /   The black Sea
(17)   162   /   Archaic paintings
(18)   169   /   The human mind
(19)   226   /   The human mind, again!
(20)   246   /   Human Nature and the World!

> > > > >

Introduction

*　　*　　*　　*　　*　　*

The other day, I was fascinated by a satellite photograph of Europe and the Mediterranean on a cloudless night, (Fig. 1). It shows all the big cities of The Old-World as glittering spots of light. The glow of each spot came from thousands of street lights, millions of homes and millions of people cooking, eating, reading, dancing, and moving around in trains and cars.

Looking at Mankind from that far away is a truly thought-provoking idea. How would a mysterious extraterrestrial visitor, looking from that high, perceive our earthly home? I believe his impression would be as follows:

> "On that blue planet, third from the Sun, there are arbitrary lines dividing it into about 190 divisions called 'States'. Oddly enough, these lines have changed, to and fro, hundreds of times throughout the last 5000 years, usually by wars in which thousands or millions of people killed each other. In every state, there are several cities, one capital, a government, a national zoo, and a parliament. For every State there is a national army whose supposed function is to patrol and protect the artificially-sanctified borders. But in many cases its real function proves to be protecting the government rather than otherwise. The total number of Mankind has exceeded six billion, and still counting. Almost half of them barely survive from one meal to another. Forests get erased and trees get felled every day to make room for thousands of new homes and new cities."

This is a bird's-eye-view of the human condition in the 21st century. If it happens to make you sad, then I believe the best place to contemplate it is to go and visit the zoo in your city. You should better avoid the week-ends and holidays, otherwise your *meditative* visit will be utterly disturbed by hundreds of little human animals called children.

On a quiet day in a zoo you will notice, maybe for the first time, the shear magnificence of such a place, full of fences, ponds and lakes. There are thousands of different animals, birds, fishes and marine life forms, snakes, snails, insects, and not to mention trees and flowers of every color and shape. An ideal zoo would also display several colored pictures of microscopic forms of life: bacteria, algae, ciliates, plankton, ameba, parasites etc.

Mankind is no more than just one single species out of millions, and oddly enough, his physical body is the weakest and most vulnerable of them all. His skin is naked, with no protective fur or fat, a few hours of cold or heat could easily kill him. He lacks the protective shell of a tortoise, the fangs of a beast, the muscles of a gorilla, the tail of a crocodile, or the wings of a bird. He can't run fast enough to escape a predator or catch a prey. Look around you in the zoo, and just think of it. How come that we, with our frail bodies, could have ruled the Earth, and locked all those animals in prisons and cages? We are made of the same stuff as they are. The genes in our cells are exactly the same in the cells of every form of life, even bacteria and worms. The genetic code of a chimpanzee is less than 1 % different from that of Hitler, Churchill or Marilyn Monroe.

Genes are the universal bricks used in building the body of every creature that ever lived. It is only very recently that we began slowly deciphering their secret code. It is becoming evident that this is the hideaway of the greatest mystery on Earth; not only of the physical structure of life, *but even beyond that.*

In this book, I have enjoyed, very much, summarizing the two main frontiers of science of today. The first is the physical structure of the Universe. We have been able to decipher its origin at a fateful moment, fifteen billion years ago. We could also follow its evolution over the eons. It was easy to notice that there is a remarkable degree of scientifically evident *'directionality'* or even *'purposefulness'*, in the evolution of the universe.

The second frontier of science is the physical structure of life, symbolized by the genes. Is there a similar degree of 'directionality' in the genetic code, comparable to that of the universe? Would it be a mere hunch, a notion, and wishful thinking? Or could there be a scientific basis for that? It was my greatest pleasure to put my finger on just that scientific basis. The implications are momentous. They are almost similar to the flood of science that followed a joyful cry *EUREKA!*, uttered by Archimedes in the year 200 BCE. He was enjoying his bathtub when the idea of the famous principle of fluid displacement flashed into his mind, and he simply ran into the street half naked and shouting eureka, eureka, which is Greek for 'I have found it!'

The word 'GENE' is so common today that you can't miss seeing it everyday in newspapers or magazines. But does everyone grasp its meaning, or its implications on the life of every human being? I doubt it. To get that kind of knowledge you may need to read a great deal of books, which you may find full of technical terms. Let me first summarize it for you in a small capsule in the first chapter. Following that, all chapters will build upon each other. All the time you will notice a faint thread of continuity binding the physical structure of atoms, universe, genes, life on Earth and Mankind until you reach a climax of wonder and intrigue on the end page.

Hussein A. Amin,
Cairo, 2005

Figure (1) Great cities of The Old-World, on a cloudless moon-lit night.

CHAPTER 1

* * * * * * * *

What is meant by

'GENES'?

* * * * * * * *

     Charlemagne was a wise ruler of medieval Europe, who encouraged schools and science. If a courtier told him that there is a Universal language used by flowers, trees, birds, snakes, fish, animals, as well as Mankind, then His Majesty would have indignantly sent the poor man into the mad-house. But it is true..!!

Since the dawn of history people have known about heredity, albeit not knowing how it works. They knew that a child inherits some of his features from mom and some other features from dad. In the old agricultural civilizations people discovered, and perfected, two main arts: the cross fertilization of plants and seeds, and the cross fertilization of domesticated birds and animals. They called it animal husbandry. The aim was to get better crops, more milk, faster horses, etc. Artificial breeding produced a type of pigeons with perfect homing-sense; people used them to transport messages. For thousands of years dogs of all varieties, shapes and purposes were created by our ancestors. We inherited their art, and all we did was give it a new name, 'Artificial Selection'. It was this very term that sparked the idea of 'Natural Selection' in the mind of Charles Darwin, later modified to become the Theory of Evolution.

All we knew fifty years ago was that plants and animals are multi-cellular organisms, and that each cell has got a smaller nucleus inside. An ameba is a unicellular organism, and still with a nucleus. Bacteria are also unicellular organisms, but without a nucleus. The only animal cells without a nucleus are the red-blood-corpuscles, which cannot divide; they are mass-manufactured by the blood-forming organs. Bacteria, which have no nuclei, can divide every few minutes without any fuss. But the division of nucleated cells is a ritualized procedure that starts in the nucleus followed by the cell-body.

Plant cells vary in size from 10 to 100 microns, (a micron is one thousandth of a millimeter), which means that the largest plant cell is one tenth of a millimeter. Ameba size is a little smaller. Animal cells are even smaller, 10 to 30 microns. The cell-wall in plants is made of cellulose, a rather hard material that forms a sort of skeleton for the plant tissue. Still, this cellulose wall is not a rigid barrier. It is an active, very alive, material that controls all the chemistry going in and out of the cell. Animal cell-walls are softer, made of carbohydrates and proteins, and are very choosy of what goes in or out of the cells.

Medical books, half a century ago, taught that the cell nucleus contains paired threads of a protein material called Desoxyribo-Nucleic-Acid (DNA). Their number in human cells is 23 pairs, which means 46 threads called chromosomes. Twenty two of them were arbitrarily numbered in approximate order of their size, from the largest (No.1) to the smallest (No. 22), but we didn't have the faintest idea of what they do. The remaining pair was the only one to which we could ascribe a function, and that is defining the sex, male or female. It consists of two large chromosomes (called XX) in women, or one X and a smaller chromosome (called Y) in men. In size, the X comes between chromosomes 7 and 8, whereas the Y is the smallest.

Outside the nucleus the rest of the cell is called its cytoplasm. It contains numerous minute structures which can be compared to kitchens, washing machines, air-conditioners, etc, because they do comparable functions inside each cell. It also contains DNA threads called mitochondria which are also related to heredity. The DNA in the nucleus comes half from mom and half from dad, but that of mitochondria in the cytoplasm comes only from the mother.

Cell division starts in the nucleus. Each chromosome makes a faithful replica of itself, and then the nucleus divides having a complete set of 23 chromosome pairs in each half. This is followed almost simultaneously by the division of the whole cell into two. Such division, (called *mitosis*), holds true in all body cells, called **somatic cells**. There is another different group of cells called **germ-line cells**. These are the ones destined to make future sex cells, ova or spermatozoa. They start to differentiate as early as the 4th week of fetal life. They migrate into the area which will form the future ovary or testis, and then almost immediately proclaim autonomy or self-rule...! They are hardly affected, from that moment on, by anything that goes on in the rest of the body. Changes in the body cells are not conveyed to the sex cells. Strictly speaking, they belong to the future generations rather than to the body which hosts them. When they divide to form ova or sperm, the DNA material is halved, so that each sex cell gets only 23 single chromosomes, such division is called *meiosis*. When in the future an ovum gets fertilized by a spermatozoon the 46 number is regained again, half from mom and half from dad. Meiosis is the technique of sexual multiplication; the embryo differs a little from its parents. But in unicellular organisms such as ameba that multiply by mitosis, the daughters and granddaughters, etc, are all exactly the same.

In 1953, a breakthrough took place in Cambridge University, Britain. Three young scientists successfully demonstrated the double-helix structure of the DNA molecule by using the technique of X-ray crystallography. They were an American biochemist, James Watson, a British molecular biologist, Francis Crick, and a British biophysicist, Maurice Wilkins. Their discovery didn't hit the newspapers at the time, but today most scientists consider it the most momentous discovery of the century, if not the millennium. A full nine years had to pass before the scientific community started to realize the magnitude of their discovery. The three of them were granted the Nobel Prize in 1962.

The chromosomes were found to be packets of a coiled thread. If unraveled, the length of the thread in each cell would reach six feet. The thread consists of two parallel spiral chains of DNA wound together in a double-helix configuration having a diameter of 2 millionth of a millimeter. The two chains wound around each other actually resemble a ladder twisted lengthwise around itself. This description gave only the 'skeleton' of DNA. The more important part, for which the Nobel Prize was offered, was deciphering the 'substance' and 'job' of each single unit on the long DNA chain.

THE SUBSTANCE OF THE DNA CHAIN

A chain is a series of objects bound together in a long line. The objects may be replicas of each other, or may be of different materials, shapes or colors. Their number may be tens, thousands, or billions. For example, the other day you gave your wife a necklace of pearls on her birthday. This is a chain, the pearls are held together by a string needled through each pearl. You also gave your daughter a golden necklace holding in the middle a heart-shaped beautiful gem. The golden necklace is also a chain, made of rings, each one binding the one before and the one after. A similar chain made of iron rings can be used for a different purpose. It can lock the door of your garden, or can be wound around the neck of a slave in a cruel sort of necklace. DNA is also a chain of organic molecules called nucleotides, but it is manufactured by Nature, and not by Mankind. There is one common characteristic for *all* chains, whether natural or artificial: once a chain is formed it can *never* be broken, except by the deliberate action of Mankind, no one else. Using a scissors, he can cut the string holding the necklace of pearls. Using a tiny flame, he can cut the rings of the golden necklace. Using certain enzymes, he can cut the DNA chain.

There is one major difference between the DNA and all other chains. Rings have only two points of 'action', arranged lengthwise. One is holding the ring before and one holding the next ring, for beads and pearls the string does the same lengthwise function. DNA chains are different. In the paired DNA chain, each ring or unit is a certain organic molecule which has got four, rather than two, points of 'action'. There are two lengthwise points of contact holding each chain intact as well as two other points of action, one on each side of the unit; they perform different jobs. (See figure 3).

On the side facing the sister chain, the job of each molecule is to faithfully guarantee an exact future replication of the whole chain, or *heredity*. On the other side, facing away from the sister chain, the job of each molecule is related to *function*. It manufactures the various proteins needed for the life of its owner, whether it is a single-celled ameba, or a multi-cellular fully grown human.

HOW IS HEREDITY PERFORMED?

The beads or units of the DNA chain are organic molecules of a chemical substance called Desoxyribo-Nucleic-Acid, abbreviated into DNA. DNA molecules are technically called nucleotides, and we will use that term from now on. In the DNA, there are four types of nucleotides, called adenine, thymine, cytosine, and guanine. The nucleotides in the two chains are matched in such a way that A (adenine), in one chain is always paired with T (thymine), in the other, while G (guanine), in one chain is always paired with C (cytosine), in the other:

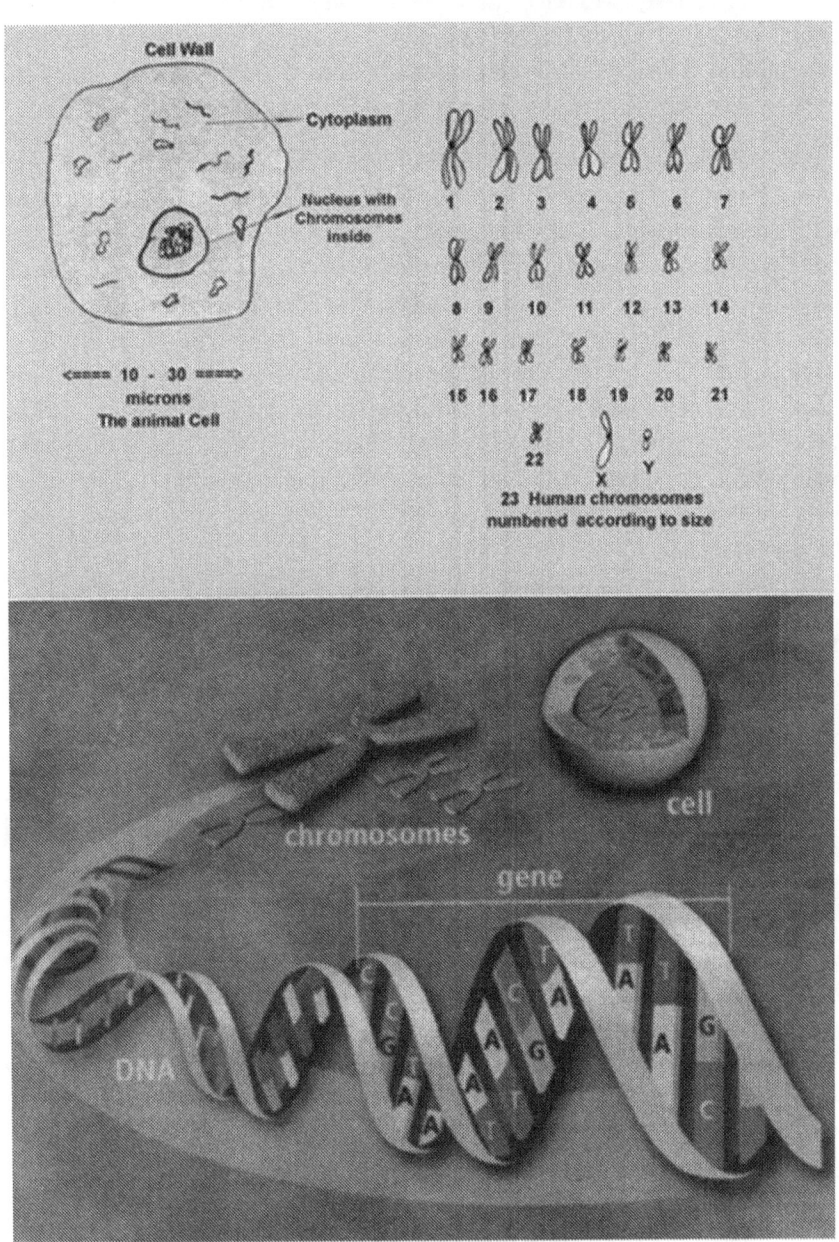

Figure (2)

A-T, T-A, C-G, G-C, and so on, along a three-billion long double chain. Thus the two chains bear a precisely complementary relationship to one another. The nucleotide sequence of either chain directly specifies the sequence of the other chain. This is what determines heredity.

When a cell decides to divide, its paired DNA chains get separated and each single chain goes to one side of the cell. Following that, each chain will act as a *'template'* for the creation of a new sister chain. There is a pool of all sorts of organic molecules in the nucleus and cytoplasm of each cell. The pool includes all four nucleotides, adenine, thymine, cytosine, and guanine. Their source is the food you eat, or the blood and milk of your mother. The nucleotides sequenced along each chain will attract their complementary *'sisters'* from the common pool. A will attract T, and vice versa, C will attract G, and vice versa. Thus, in no time, the double helix will get fully and faithfully replicated again. Each daughter cell will get a fully formed double helix, an exact reprint of the one which was in their mother cell.

This is simply the miracle of heredity!

HOW DOES THE DNA CHAIN GOVERN CELLULAR FUNCTION?

For each nucleotide, the fourth point of action is on its outer free side, facing away from the sister chain. It is exposed to the pool of organic molecules in the nucleus, ready to interact in order to fulfill any needed function.

It was very interesting to discover that only about three percent of the whole sequence of the DNA chains, (on their outer free side), do share in any function. The remaining bulk of 97 % of the millions of nucleotides is *apparently* doing nothing at all, and has been termed as *Junk-DNA*. The functioning three percent are not in the form of a continuous stretch. They are present as segments of various lengths and numbers. Each segment may be formed of a few hundred to a few thousand nucleotides. Every such functioning segment is called a Gene. (See figure 3.)

In the nucleus of the human cell the total number of sequenced nucleotides is about three billion; this total is what we call the human 'genome'. Out of these, there are only about 90 million functioning nucleotides, or only three percent of the total. These are subdivided into a total of about 100,000 genes, irregularly distributed among the long stretches of Junk-DNA. However, recent research in 2003 suggested a smaller number of only 30,000 human genes. So, we say that humans have about 30,000 genes. A round worm has about 19,000 genes. A fruit fly has only about 13,000 genes. A bacterial cell has no more than 3,000 genes.

HOW DOES THE GENE FUNCTION?

Proteins constitute about   80 % of the dry weight of muscle, 70 % of that of skin, and 80 % of that of blood. The inside of a plant cell is also full of proteins. The importance of proteins is related more to their function than to their amount in an organism or tissue. Examples are hormones, neuro-transmitters, anti-bodies, hemoglobin, chlorophyll, skin pigments, digestive and other enzymes, etc. Almost every active substance in living cells is made of proteins. The function of any cell is determined by the specific protein it creates.

Proteins are composed of chains of various combinations of simpler compounds called amino-acids, repeated many times, and strung together in a particular order for every particular protein. Amino-acids are the building blocks of proteins. *There are only twenty known amino-acids.*

The sequence of amino-acids in all proteins is genetically determined by the sequence of nucleotides in the cellular DNA. When a particular protein is needed, the DNA code for that protein gets transcribed into a sequence of nucleotides along a segment of RNA. The RNA segment then serves as a template for protein manufacture.

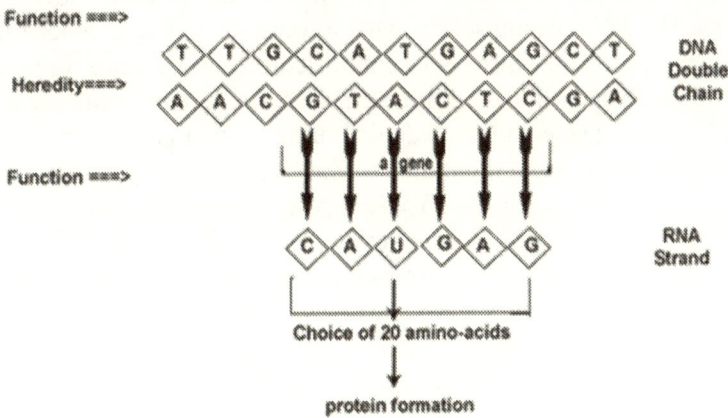

Figure (3), Diagram of DNA and RNA,
Showing how they fulfill both heredity and function.

WHAT IS RNA?

Chemically it is called ribo-nucleic-cid, abbreviated into RNA. It is a cousin of DNA, and acts as its messenger, or secretary. Unlike DNA, RNA consists of single, rather than *double* strands. But like DNA, each RNA strand is formed of similar nucleotides, adenine, guanine, cytosine, but the fourth is uracil instead of thymine. It does not exist naturally like DNA, but gets assembled whenever the DNA wants to send a message outside the nucleus. After it has carried out the order, the strand simply disassembles itself.

Let us take an example of a liver cell. Its nucleus has two DNA strands, each having 3 billion nucleotides. Their opposing faces are concerned with replicating the cell if it needs to divide. Their far faces have about 30,000 active segments called genes. When it wants to manufacture a certain hormone, which is simply a protein, it will switch-on the gene coded for that protein. In the nucleus the codes are those of DNA, consisting of four symbols, namely A, T, C, and G. The switched-on gene will start collecting nucleotides from the pool inside the nucleus. C attracts a G, and vice versa. T attracts an A, but A attracts a U. Uracil is the replacement of thymine in the lowly language of messenger RNA. The resultant RNA strand consists of a special sequence of A, U, C, and G. It leaves the nucleus into the cytoplasm of the cell in order to fulfill its orders.

In the cytoplasm, it starts collecting amino-acids from the pool of all sorts of organic molecules. Amino-acids are the building blocks of all proteins. The RNA strand selects what it wants according to a beautifully coded language. Every three adjacent nucleotides constitute a unit known as a **codon**. Each codon is simply a code for a certain amino-acid. For example, the sequence AUG, (adenine, uracil, guanine), is a codon that specifies the amino-acid methionine. There are only 64 possible codons, nick-named as 'RNA triplets'. You can test it yourself on any computer, just try making all possible three-some combinations out of four letters. Three codons do not code for amino-acids, but indicate the end of a protein. The remaining 61 codons specify the twenty amino-acids that make up all proteins. The AUG codon, (in addition to coding for methionine), is found in the beginning of every RNA strand and indicates the start of a protein.

Proteins are similar in all forms of life, from bacteria and ameba up to humans. Hence, the DNA nucleotide sequence (the gene), that creates any particular protein is exactly the same in all forms of life. Today, we are able to play around with this idea. We can pick up a certain gene from a plant or bacteria and stick it into another plant or animal, and it will do its same function, perfectly well. It will make the same protein it has been originally coded for.

LANGUAGE OF THE GENES

This is simply the miracle of the genetic code. A secret code of four letters, plus a fifth for RNA, that gets translated into a language of twenty alphabetical letters (20 amino-acids). The code and language are universal for all forms of life, and have been the same since life began 3,500 million years ago. As we have seen, this miraculously arranged genetic code determines both heredity and function. But there is one major remaining question. If only about three percent of the DNA sequence in any living organism is functionally active as genes, then what is the bulk of 97 % doing? Why does the living cell keep eternally replicating them if they have nothing to do? Until now scientists have called them Junk-DNA, but they are not junk at all. In a later chapter, the Gene-Suppression-Activation-Theory will examine and explain the purpose of this so-called Junk-DNA.

THE GENES AS DIFFERENT FROM DNA

DNA is an organic substance. But the 'gene' is an information code, expressed in the form of a certain sequence of nucleotides along the DNA chain. DNA is your computer chip, the material hardware. But the gene is the intelligent information that you put on that chip.

They are not synonymous. Genes are encoded information, quite distinct from the chemical medium on which the information is recorded. It is the same as the information conveyed in a book, being quite different from the ink and paper on which it is printed. Now, the overwhelming question remains: *Who initiated this encoded information? Was it Nature, through haphazard physical means? Or was it a willful intelligent force called God?* This is the trillion dollar question!

GENE-SUPPRESSION

Embryonic life starts at the moment of fertilization. At first there is a single cell, then 2, then 4. Each of them contains the whole set of 30,000 genes fully active. You can compare them to a huge encyclopedia telling each cell how to make a complete human being, hence the occasional occurrence of identical twins. We call such cells as being *totipotent*, capable of reading and obeying all the pages of the encyclopedia. However, starting from the third division, when the number of cells is eight, specialization or differentiation, will start. Some cells will make the skin and nervous system; we call them **ectodermal stem cells**. Others will make the heart, vessels, muscles, bones etc; we call them **mesodermal stem cells**. A third group will make the gastrointestinal tract, liver, pancreas, etc. We call them **endodermal stem cells**. The capabilities of stem cells are limited to any of the mentioned three groups of organs and tissues. They cannot make a whole organism; hence they are termed as *multipotent*, and not *totipotent*.

In the animal kingdom the only totipotent cells are the first eight ones, which get formed by the first three divisions of the fertilized egg. Hence, the number of twins in humans can never exceed eight. However, in plants virtually every cell is totipotent; you can create a whole plant from a mere cutting of a branch, a leaf or even a single cell. This is a sort of asexual multiplication. But still a plant can multiply using its seeds as another form of multiplication, sexual multiplication, same as the animal kingdom. Plant life combines both means of multiplication.

If we get a cell from the stomach wall, and another from the heart muscle and examine them, we will find that each contains exactly all the three billion-lettered genome, including all the 30,000 genes. The encyclopedia is present in each, but the stomach cell will read and obey only the pages that tell it how a stomach cell should behave. It will *somehow* suppress all the rest. If a stomach cell would be able to read and obey all pages it would make a whole human being, rather than function as a stomach cell. It is the same also with cells from the heart, the liver, the skin, the brain, etc.

On reaching maturity, cellular multiplication ceases in some organs such as the neurons and renal cells. Other cells may retain their ability to multiply for the rest of their owner's life, such as skin cells, the lining of the gut, the lining of blood vessels, bone cells, and blood cells including those of the immunity system, and sperm-precursors. Liver cells also can occasionally multiply according to need. But in all these cases the going-on multiplication is always under strict limits and rules, otherwise it could become pathological. If you get a skin wound, it starts to heal by multiplication of skin cells on both sides. Meanwhile, each cell is in continuous communication with its neighbors, right and left, above and below, as to the direction, the shape, the color, as well as when to stop.

This is one of the great secrets of life. How a gene gets suppressed, switched-off and / or reactivated? How does a cell manage to label some of its genes as temporarily-out-of-order? We are still far away from deciphering how each gene is jump-started to activity or inactivity. This is a mystery, even much more intriguing than the discovery of the genetic code itself.

GENOMICS

Some years ago the U.S. president and the British prime minister boastfully told a press conference that their research laboratories had finished recording the human genome. It looked as if it were a breakthrough. But it was just a public-relations fanfare, meant to inaugurate a new trade, mockingly called 'genomics', like 'economics'. What their laboratories really did was just list the 3 billion letters of the genome, TCGATCA, etc, nothing more. The real breakthrough took place in 1953, in deciphering the genetic code, for which Watson, Crick, and Wilkins earned their Nobel Prize.

The next huge task confronting workers in the genome project will be called proteomics, says John Richards of the California Institute of Technology; [Newsweek, Feb. 19/2001, page 54.] The genome is a set of instructions for making proteins. Yet, from a single gene you can get ten or more different proteins, a genome analysis alone won't tell you which ones. There are probably half to one million human proteins, from hemoglobin, insulin, brain chemicals, sex hormones, to countless enzymes that keep the cells running.

Now let us get some insight into this potentially multi-billion dollars trade of genomics. The techniques used include many technical terms such as: shredding, freezing, splicing, amplifying, tagging with special dyes visible under laser light, sequencing, assembling, etc. For our purpose we are interested in the results rather than the techniques. I'll choose only some of the items that may interest the general population.

(A) STUDY OF THE AGING PROCESS, AS WELL AS POTENTIAL LINES OF CANCER CURE.

It was discovered that at the end of each chromosome there occurs a repeated stretch of meaningless 'text': the 'word' TTAGGG repeated again and again about two thousand times. This segment is known as a 'Telomere'. Its presence enables the DNA-copying devices to get started without cutting short any sense-containing text. Like the little plastic bits on the ends of a shoelace, it stops the ends of the chromosome from fraying. The phrase TTAGGG is the one used in the telomeres of all animals, protozoa and fungi. In the plant kingdom, the phrase has an extra T at the beginning reading TTTAGGG. Suppose a photocopying machine has got the bad habit of missing a line or two at the beginning and end of every page, then by the time a multi-page document has been copied 20 or 30 times lots of information will be missing. To avoid this, you would leave a blank space at each end. This is the exact idea of the telomere.

An enzyme called telomerase was discovered, which repairs the telomere ends of all chromosomes during the first few totipotent divisions. But once specialization starts telomerase activity stops, and the telomeres start to lose some length with every single division. When the telomeres get lost the chromosomes will start losing some of their genes with every division. Cell quality and life quality will deteriorate. Thus, telomeres are the count-down of ageing. They are similar to a stop-watch which gets wound early in life, setting a limit to the lifespan of every animal or plant. In dogs the average lifespan is 10 to 12 years, in the horse it is 30 to 40 years, in Mankind it is 70 to 80 years. Modern health care may increase the average human life-span, but the recorded limit never exceeded 120 years.

Interestingly, it was found that in tumor cells the telomerase enzyme is frantically active. It keeps on repairing the telomeres of the malignantly dividing cells, thus removing any obstacle against their runaway growth. This discovery opened the way for research on the possibility of finding a way to stop the telomerase enzyme, as one potential line of cancer therapy. Interestingly also, it was noticed that Dolly, the cloned sheep, had lots of medical problems related to ageing such as obesity and osteo-arthritis, in spite of being one year old. Dolly was cloned using the nucleus of an adult sheep. This shows that her DNA was already ageing even as a newborn. This then will always be a problem for the cloning business.

(B) PREDICTION OF GENE-RELATED DISEASES

We can take a cell from a newborn, and study its genome. If we find certain genes, then we can almost surely predict that the child will develop a given disease at a given age. If we study a fetal cell in early pregnancy, we can present this information to the parents and their doctor to make decisions. If we advance a further step, and study the genomes of the parents, we may be able to predict certain diseases in their offspring.

Potential ethical problems will surely arise if the use of genome studies becomes a routine test. For example, by studying the genome of a young undergraduate we may be able to foretell quite a bit of his future health. When he applies for a job, the recruiting company may ask to see his genome study, and could refuse his application. An insurance company may demand the same, and could deny him an insurance policy. This route of genetic application is a very hazardous one, full of political and social implications, especially if it gets stored on computers. People would lose all privacy, as if standing naked on the internet screens..!

(C) GENETIC FINGER-PRINTING

In the good-old-days many a criminal could get away with a crime because it was difficult to pin-point his guilt scientifically. Blood groups and fingerprints were the only tools of forensic medicine. Nowadays we have potentially fool-proof techniques. Researchers noticed that in the intervals between genes, (in what is called Junk-DNA), the nucleotide sequencing shows minor differences from one person to another. The differences are permanent in every cell of every person and are easily detected. This can prove the identity of any individual even if he had plastic surgery to change his face. Still, the classic fingerprint is quite unique. You may be surprised to know that identical twins have the same DNA, but they do have *different* fingerprints.

(D) *SOMATIC* GENETIC THERAPY

In somatic genetic therapy, a patient is treated by changing his genes, instead of giving him a medicine or irradiation. The word 'somatic' implies that in this type of therapy we are dealing with the body cells, and not the germ-line cells. Whatever genetic changes we make will affect the individual alone, and will not get transferred to his offspring.

First, we must explain something about the similarity of the genes in all forms of life, or in other words *the universality of the genes.* Cellular life depends on protein molecules. These are manufactured in the cytoplasm by the DNA - RNA mechanism, which uses a four-lettered code of nucleotides and a twenty-lettered alphabet of amino acids. This is universal. Hence it is not strange to find exactly similar genes in widely different forms of life. When we find that the nucleotide sequence of a known human gene is the same as a gene we find in a mouse, a bacterium, a plant, an insect, or a fish, we must deduce that it performs exactly the same function in all of their bodies. For example, a gene in E.Coli bacteria is exactly similar to the gene which manufactures starch in the potato plant. It manufactures starch for the metabolism of those bacteria. This knowledge led researchers to try transferring the gene from the bacteria into the potato plant. This genetically engineered potato produced 20 % more starch than ordinary potato. This increases its energy value, and also reduces its water content and makes better crisps, chips and French fries.

Genetic research laboratories have been busy decoding the genome of the Drosophila fly in particular. Why do these scientists bother? Because of about 300 identified human-disease genes, there are 177 of them which have analogs in the genes of that fly, including 68 % of cancer genes. A worm called "C. elegans" has analogs of almost half of the recognized human-disease genes. The usefulness of genetic decoding is clear. Suppose we get a patient who lacks a certain gene, and hence gets a certain disease. We could search for that gene in any life form. A useful gene can be 'spliced' into a virus after removing its DNA, and then introduced into the body, where it enters into body cells, practically introducing the gene as well. The added genes in these techniques go nowhere near the germ cells that will form the next generation. That worry has been firmly laid to rest. [Ridley, 1999]

In Scientific American magazine, [Feb. 2003, page 16], there is a report about a gene in fish that causes regeneration of its heart muscle. An exactly similar gene is present in humans. This observation could lead to strategies for scar-free repair of the human hearts. This field is offering great potentials for future medicine. But it is still experimental to a great extent. Patients must accept that they are taking calculated risks.

PHARMACO-GENETICS

Pharmaco-genetics is another field of genetic medicine. This area of medicine came about when it was noticed that the effect of the same medicine may differ from one patient to another, simply because his genes which deal with the effects of this medicine are different from other people. In this new field we can tailor the medicine to the needs of our patient's genes, giving him a better result. Some valuable human proteins, such as growth hormone, insulin, blood clotting factors, etc, can now be commercially manufactured. The field of pharmaco-genetics is expanding rapidly. Scientists can simply introduce a gene into the genome of an animal, engineered in such a way that the desired protein gets secreted in the animal's milk, and then harvest it with utmost ease. Similarly, we can introduce genes into certain bacteria, and convert them into virtual factories for the synthesis of certain proteins.

(E) GENETIC ENGINEERING

In this discipline scientists deal with the germ-line cells that belong to the future generations. First, it is worthwhile to notice that we are not creating life, or creating anything at all. We are just exploiting the existing processes of Nature, albeit processes which are still mysterious and arcane.

In the world of plants, genetic engineering is a widely used industry. Our ancestors used it in the fields by cross fertilizations of various plants to get better crops. Today, what we do is almost the same, but on a more scientific basis. We do it in the labs rather than in the fields. We can create tomatoes with a longer shelf-life; wheat that doubles the crop per acre, cotton leaves that are automatically resistant to the infamous cotton worms, instead of using insecticides, etc. Plant cells are totipotent, so if we change a gene in any single plant, it becomes as if it is a new species created overnight. It will propagate that gene into the seeds in a permanent way.

This is a very costly research, so the companies which invest in it will always try to reap billions of dollars in return. They call it 'Intellectual Property' that must be protected by law. Sometimes they deliberately engineer their seeds in certain ways, so that the farmer will not be able to replant it again, neither by using plant-cuttings nor by the resultant seeds. The seeds are deliberately made sterile. A peasant will have to buy *new* seeds every season. Europeans are still wary of genetically engineered foods, but Americans are not. They say that plants have been engineered by Mankind for thousands of years. Doing it in the laboratory rather than the field is exactly the same concept, and shouldn't cause any worry.

Genetic engineering in the world of animals is a totally different story. It has to start before fertilization. Here we deal with sex cells, rather than body cells. Thus the changes we make will surely get relayed into all future generations. We can make cattle and chicken produce better quality meat or milk, better resistance to germs or parasites, etc. We can introduce human genes into mice or pigs, so as to get tissues and organs that could be transplanted to man without rejection, etc.

Genetic engineering research in the world of Mankind is forcefully resisted by governments, non-governmental organizations, and the Church. Our germ line sex cells are not our property; they belong to the future Mankind. We dare not doom them into changes about which we hardly know enough. Even the slightest tinkering with human genes may give un-foreseeable changes in the qualities of Mankind in 30 or 40 generations. Are these changes for the better or the worse? Nobody can dare to guess.

(F) STEM-CELL RESEARCH

Stem cells start forming at the fourth division of a fertilized egg. From that moment on, cell division will create three types of stem cells: ectodermal, mesodermal, and endodermal. They are multi-potent; a mesodermal stem cell for example can produce blood vessels, lymphatics, heart muscle, skeletal muscle, bones, cartilage, etc. If we can harvest a mesodermal stem cell from a patient, and then multiply them as tissue cultures, then we may be able to get any of these tissues and graft it into his body without fear of any rejection. Another source for mesodermal stem cells is the blood collected from the umbilical cords and placentas. It contains stem cells that can rebuild the blood and immune systems of people with leukemia and other cancers.

Few stem cells can be harvested from adult bodies. So the research has turned to fetal cells. One evident source is aborted fetuses. Another is from frozen discarded embryos during in-vitro fertilization. A third source is deliberately created embryos using the cloning techniques. This line of research is burdened with serious moral implications. So much so that it will rule national and international debates for years to come.

(G) FREE-WILL – VERSUS – GENETIC DETERMINISM

Genetic determinism means that the life, as well as the behavior, of the living creature is totally governed by its genes. Plant and animal lives are totally under the rule of their genes. Genes dictate to every animal how to feed, how to procreate and how to raise its offspring just until it can run or fly. Beyond that, the offspring completely relies on its own set of genes. Mankind is not like that, we are different. Sometimes we can defy our own genes. We do that every time we use contraceptives, or whenever we go on a hunger strike for a noble cause. However, this debate is as old as humanity. It deserves a little detail.

People who claim that Mankind is also under genetic-determinism often desperately try to extend its significance into the everyday morals and norms. A lawyer could argue that his serial-killer client is innocent, his crimes were committed *not by him*, but by his genes. The lawyer may ask for a not-guilty verdict. Homosexuals argue that it is not their fault to be different from the mainstream, because *'it is all in the genes'*. Hence they demand legitimacy, equal rights, and even representation in legislative bodies. All this is not true; it is a sort of wishful thinking by people who, deep in their souls, wish to defend themselves against what the rest of society considers as deviant behavior. A recent title of a scientific article in Newsweek reads as such: ***"No, It Is Not All In The Genes!"*** [Newsweek April 10 / 2000, page 61, by Robert Sapolsky Professor of Biological Sciences and Neurology at Stanford University]. He calls it *Genetic-vulnerability*, but not *inevitability*. A similar theme is reported in TIME magazine, June 2nd, 2003, page 51: *"What Makes You Who You Are"*, by Matt Ridley.

Much of the DNA simply constitutes on and off switches for regulating the activity of the genes. It is like you having a 100-page book, and 95 of the pages are merely instructions and advice on how to read the other five pages. You can't dissociate genes from the environment that turns genes on and off, *and* you can't dissociate the effects of the genes from the environment in which proteins exert their effects. ***Genes are not the whole story***...In this regard, the following example is illustrative: [Watson, J., 2004, page 403]: He says let us suppose there is a particular plant species in which height varies genetically. Let us put one set of seeds in a tray with high grade soil, and another set of seeds in a tray of poor soil. The environment, (in the form of soil quality), will surely affect the plants. While genetics is the dominant factor in determining height differences among plants within a tray, genetics has nothing to do with the differences between the soils (environment) of various trays.

Scientific behavioral research has clearly shown that *nurture*, by environment, society, and family, gives more than 80 % of the end-result character of any man or woman. What the genes provide for a newborn is only a background of extrovert or introvert personality, an aggressive or non-aggressive temperament, etc. All the rest of the child's character gets shaped by his nurture, and especially so by the sort of intimate relationship he had with his parents, particularly the mother. Developmental psychologist Jerome Kagan of Harvard University reported as long ago as the late seventies that shyness has a strong genetic component. In spite of that, he found that if parents provide extra nurturing and a gentle but firm guiding hand, by the age of four most previously inhibited children overcome their "innate" shyness. [NESWEEK, special edition titled "Your Child" fall /winter 2000, page 60.] We may indeed come ready-made from the factory, almost like a television set. But many of our genes seem to operate like volume controls. Good parenting and nurture can turn those volume controls up or down. In the field of raising children, we are not slaves to our genes. Hence, Western societies are now starting to ponder the excessively harmful effect of many child-rearing habits, for example the literally absentee *working* mother.

By definition, a gene for homosexuality, if it exists, would be a sterile one. It won't be transferred to offspring, *because there will be no offspring.* Sometimes homosexuality is the only sexual outlet available in isolated unisex groups such as boarding schools, nuns, celibate clergy, prisons, wars, and nomad life. Often the reason for homosexuality starts by the first exposure of the child to sexual knowledge, especially if it was by an ignorant peer or a malevolent caretaker. Sometimes parents could carelessly act with a small child in the room, believing that the child couldn't even notice. This is wrong, and if the child identifies with, and loves, one parent more than the other then seeds of homosexuality can get sown. The trauma of sexual abuse in childhood, especially by an adult of the same gender, may also initiate homosexual inclinations in later life. A third uncommon cause of homosexuality takes place during pregnancy. If the mother is given hormonal medications, this could sensitize the fetal brain into the wrong sexual orientation. Sexual orientation of the brain starts during fetal life by the effect of the fetal hormones on its own developing brain. External hormones can alter this balance. Pregnant mothers should be warned. They should also constrain themselves in ways of smoking, drinking, and the abuse of drugs, that could cause hormonal disturbance.

GENETIC DETERMINISM - VERSUS - RELIGIOUS THOUGHT

The debate here is about the accountability of Mankind for whatever sins they commit. Understandably, part of the debate is about supernatural accountability. This is the realm of philosophy rather than science and I believe that philosophers will have to ponder these questions from now till eternity. There is a new approach to exploring brain structure and function using recent and very sophisticated techniques. For the first time it looks possible to address questions about spirituality in a scientific way. This is a promising, and most intriguing discipline, about which I'll give more detail in chapters 5 and 11.

CHAPTER 2

* * * * * * * * *

Who is Darwin?
and what exactly is
the theory of evolution?

* * * * * * * * *

Everybody became more *Darwinian* than Darwin himself. His name became the symbol of many claims which never crossed his mind in the year 1859...!

Figure (4), Charles Darwin

A few days ago, we had an afternoon party on the beach. The sky was blue, the sea was blue, and most of the swimsuits were blue. The only thing that was not blue was everybody's mood! After lunch, the ladies retired to their rooms, and the men stayed for tea, coffee and naughty jokes. There was a joke about a gorilla, and a friend who has a Doctorate of Economics confessed that he has never understood what exactly Darwin's theory is. All he knew was that at one time Mankind were monkeys. His confession was mirrored by some others, in spite of their highest professions. They asked for half an hour of general knowledge, and it was a pleasure for me to explain.

FIRST, WHO IS DARWIN?

Born 1809, his father was a famous London physician. At 16 he tried for two years to study medicine, but he couldn't continue and shifted to Church studies. He was fascinated by botanic observations, and collecting various types of beetles and butterflies, pigeon breeding, etc. Soon this hobby became an obsession, and he became famous as a naturalist. This hobby was uncommon among young men in his age, and it soon earned him the chance of his life. The Victorian British Empire was sending research missions to the four corners of the world. He was offered passage on a voyage around the world on a ship called The Beagle. With his aborted medical study and naturalist hobby, he was the perfect fit, being young, unmarried, jobless and from a rich family in no need of any wages. The Beagle sailed in 1831, and he was back in London five years later. It started with Brazil, and then around the coast of South America, with many stops, including a famous one at the Galapagos Islands. He then sailed to Australia and New Zealand, on to the Indian Ocean, around Africa, and finally back to Britain in 1836. All the time he was busy writing his naturalist and geological observations, as well as collecting specimens of everything. It was a great scientific wealth, well-earned, and in the very good hands of a very intelligent man.

He got married in 1839, and sired ten children.

The key observation in his research was the fact that species were *not* eternally immutable, they changed over the years. This was contrary to the common belief, (based on the stories of Creation in the Biblical Book of Genesis), that living things are, today, the same as when they were created. He performed lots of research, collected many specimens, and wrote endless notes. In the mid-19th century Darwin's famous theory was already in his mind, as scattered notes, but he hesitated to publish it. Atheistic ideas were abounding at that time, mainly as a reaction to the dogmatism and cruelty of the Roman Catholic Church. He knew that if he published his results it would fuel the claims of the atheists, claims with which he wasn't really committed. For example, Darwin supported the argument that God created matter and energy at the beginning of the universe, with a propensity not to stay put as they are, but to organize and evolve spontaneously in an endless way. He gave the name of Natural Selection to the process by which the living things gradually changed in order to better suit their habitat. The term got translated into the more popular one 'Survival-of-the-fittest'. At that time nobody knew anything about genes or how heredity took place.

One of his very insightful observations was as follows:

"Look at man's pink skin and puny frame -- how could selection evolve greater weakness without the counter-balancing gift of *Reason* being bestowed *first*? Man must have had the human proportions of mind *before* he could afford to lose the bestial proportions of body."

In 1859 he finally decided to publish his observations in his famous book 'Origin of Species'. He thought the book was too dry and perplexing, and did not expect it to sell. The publisher also agreed, and published only five hundred copies. Soon, everybody became more *Darwinian* than Darwin himself. The German philosopher Ernst Haeckel applied the term of *'survival of the fittest'* to human social life, as a support to the racial supremacy claims of Teutonic Germans. T.H. Huxley raised the banner of atheism or at least agnosticism, a term which he personally coined. Darwin's name became the symbol of many claims which never crossed his mind in 1859. His health rapidly failed in the 1870s, and he died in 1882. In spite of the atheistic row raised by his theory, he was given a very dignified burial in Westminster Abbey, as a great man of the British Empire.

It is interesting here to repeat an observation by Daniel Dennett [Dennett, D., 1995, page 44]: *"Controversy about the mechanisms and principles of speciation still persists till today, so in one sense neither Darwin, nor any subsequent Darwinian, has explained the Origin of species. The way in which species begin is still a mystery, till today."*

The naturalistic hobbies of Darwin made him very familiar with the methods of selective breeding of plants, flowers, pigeons, cattle and dogs. Dog varieties in particular ignited his curiosity. The playful cute Chihuahua can be easily hidden in a lady's handbag. The Great Dane can weigh up to 75 kilograms, and can run faster than a horse. Yet both are one and the same animal, selectively bred by Man, from an original stock of wolves. If Man can do this in a few thousand years, guided by his own desires, can't Nature do similar changes in millions of years, guided by purely *Natural* means? This was the first question ringing in his ears when he boarded The Beagle. The answer he discovered was YES!!

During his trip he noticed three natural forces which can do just that, namely: the available type of food, the presence of other animals as prey or predator, and the climate and natural catastrophes such as floods or fire. The Galapagos Islands, off the coast of Ecuador, South America, were particularly important in his observations because all the animals and birds there were geographically isolated for thousands of years. There were many small singing birds called finches that looked very similar. However, his keen observant eyes noticed lots of differences among them, so much so that he divided them into 13 varieties. Unfortunately he made a mistake of calling them 13 *species* and I believe that this wording mistake has caused much more trouble than anybody can imagine, and let me explain. Darwin detected three main differences in the finches: the length and shape of the beak, color differences of the face and feathers, and the tune of the male song. Finches are seed eaters, seeds come from plants, and plants differ from one season to another according to rain, drought, as well as the type of soil. On different corners of the islands he noticed that the shape of the beak fits exactly with the locally available type of seed. Darwin concluded that birds with unsuitable beaks will not have enough food, and will die young before having offspring. Only those with the slightly different beak will be able to eat well, and raise a family. Soon they will out-number all other types in this particular locality. Coincidentally, he noticed that each beak-configuration goes with a certain configuration of feather colors. Similarly Darwin concluded that color differences help the male and female birds select each other, so that their chick will inherit the suitable beak, and thus get enough food to survive.

Finally Darwin noticed that each finch variety had a different song. It came to be realized that singing is a genetically imprinted capacity in the brain of every singing bird. But the tune is learned by imitating its parents and siblings. In a way, this is exactly like our *genetic* language syntax and *acquired* vocabulary. If a chick gets adopted by a mother of a different type, it will *see* the wrong feather colors, and will *learn* the wrong tune. The result is that it will always get shunned by potential mates in both varieties, and will rarely be able to raise a family. Thus, Darwin concluded that each specialized male tune will attract a female of the same type. This is a still further guarantee of Selective Natural Breeding. The end result is a subpopulation of the same animal or bird that is reproductively-isolated from the rest, by purely natural mechanisms. Unfortunately the definition of the term *species* came to include, (among other things), reproductive isolation. Of course you cannot cross a fish with a tree, or a flower, or a cat, or a bird. They are reproductively isolated. It is the same term used for the finches, (same species), and big cats, (also same species). Thus, reproductive isolation per-se should not be the *only* definition of species. It must include as well similarity of appearance, similarity of behavior, and occasional variability within limits; such as the case with the finches. When we consider asexually reproducing organisms, and organisms that fertilize themselves, we base the definition of species on the degree of morphological and physiological resemblance. Under these conditions a species is defined as a group of individuals having a common number of distinctive characteristics.

It has now become possible to pinpoint the cause of the biggest confusion within the theory of evolution. A more modern term for natural selection is *micro-evolution,* within the same species. This should never get mixed with *macro-evolution.* The last term means the theoretical possibility of a fish turning into a dinosaur, into a bird, into a whale, into a gazelle, into a lion, into a monkey, into Man; either directly or indirectly over millions of years. To sum up, the mechanisms by which *micro-evolution* exerts its action are mainly three: The first is weeding-out unfit individuals, who would die young before producing offspring. The second technique is female-selection, the female of the bird or animal becomes choosy. She looks for the strongest male, or the one who advertises the best signs of good health, such as extravagant plumage. Other males are denied the chance to spread their seed. The third mechanism is mate-repugnance, where males and females reject individuals who have different beaks, different tufts of hair, different color arrangement, or different tones and pitches of song. By this mechanism the improved qualities of the species are prevented from reverting back. Natural Selection however can work only on the variables already present in the heredity genes of the animal. In this frame it has no contradiction with Creationists' arguments at all, and is really not much different from our own methods of artificial selection and breeding techniques.

It may be worth pointing out here a fact not known to Darwin at his time. The 13 varieties of finches can be *artificially* cross-bred, in spite of them refusing to interbreed in the wild nature. Similarly also, the big cats all belong to one family. Artificial cross breeding has been found possible between lions, tigers, jaguars, pumas, cheetahs, etc. But none of them would *naturally* interbreed. This shows that all these carnivores are variants of one original species, changed by natural selection according to various habitats.

Natural selection is a definitely predictable process. Thousands and millions of species have become extinct over the eons. In most cases our present scientific knowledge has allowed us to identify the factors involved in those ancient extinctions. Similarly also when environmentalists of today recite the names of all the species that are endangered by the ruthless actions of Mankind, they are actually applying the known predictable rules of Natural Selection, and nothing more. The scale of time involved in evolutionary sciences does not allow us to carry out experimental work, neither on Natural Selection nor on the theory of evolution. We only deduce our knowledge from thousands of tiny bits of scientific data that fall together like the bits and pieces of a giant jigsaw puzzle.

Yet, we have had some opportunities for an actual real-time scientific experiment in this field on the same Galapagos Islands which sparked the idea of Natural Selection in Darwin's mind. In 1973 two young researchers, Peter and Rosemary Grant, started a 20 year scientific monitoring of generation after generation of finches on one of these islands. They confirmed all the observations of Darwin, and added further modern techniques such as molecular and DNA changes throughout the study. It was published in 1994, in a book titled *"The Beak of the Finch"*.

Another modern-time proof of the process of natural selection came from the countryside of Britain. The B. Betularia is a moth that thrives in the English woodlands. It used to have light colored wings until the middle of the 19th century. Its lighter color fit well with the light color of the tree trunks on which it usually landed. Birds' eyes couldn't easily spot them, and many of them survived. In 1848 the first specimen with a dark colored wing was detected in Manchester, and shortly afterwards in other industrial regions. The explanation was another beautiful example of how

Natural Selection acts. With the advent of industry and the multiplication of soot belching chimneys the pollution changed the countryside. The tree trunks and vegetations became darker colored, sometimes almost black. A light colored moth became painfully visible to any hungry bird. Only those butterflies which were a bit darker could survive. The genes for both dark and blond colors are present in every moth. What Natural Selection did was to favor those who inherited one color at the expense of the other, according to this environmental change. No other process is involved, and particularly there is no suggestion of any *mutation* or genetic change.

Similar also is the story of resistance of bacteria to antibiotics, and resistance of pests to insecticides. The term *mutation* was coined in the 1950s, and that was in relation to how bacteria and insects acquire resistance to new antibiotics or insecticides. The term was wrongly referring to changes in the genes. In the 1950s, genetic knowledge was still in its infancy. Some said that the bacteria or insects develop antibodies against the new agents, the same way we develop antibodies against measles. Others *correctly* said that the new antibiotic, (or insecticide), kills all the bacteria (or insects), with the exception of those which *happen* to possess naturally pre-existing resistance to them. These then multiply, and *are* resistant to that antibiotic or pesticide. And so it is just another manifestation of Natural Selection, rather than genetic mutation or antibody formation. A further confirmation of this view is the famous bacterium, the throat streptococcus hemolyticus. It has never been able to develop penicillin resistance until today. This simply means that since eternity none of its colonies has ever had any innate penicillin resistance. So, penicillin will always be effective against that organism. This is a blessing for the millions of children who need long-term prophylactic penicillin to prevent rheumatic heart and kidney complications. Evolutionists are mistaken to include these stories as examples of the mechanism of evolution. When bacteria acquire resistance to various antibiotics they do not change into another species. The diphtheria bacillus does not become a cholera vibrio, nor does a tuberculosis organism change into typhoid or plague.

In my readings I also came upon a very funny proof of how natural selection works. For poachers in the jungles of Africa, elephants with big tusks are prime targets. They would ignore elephants which have smaller tusks as not worth the trouble. Over the years, hunters gradually noticed less and less of big-tusked animals and more and more of small or even tusk-less elephants. Local sages claimed that the elephants are deliberately cutting down the size of their tusks as a punishment for the poachers. But it is quite evidently another manifestation of natural selection. The ones with the big tusks couldn't survive to reproductive success. They got killed before spreading their genes. Small-tusked elephants had a reproductive advantage, with no mutations or genetic change.

> > > > >

Now let us talk a little about the other limb of the theory of evolution, which is *speciation*, new species formation, or *macro-evolution*. The theory claims that new species arise as the end result of natural selection weeding off environmentally unsuccessful creatures. It also claims that the source of the *variables* on which natural selection acts is simply haphazard genetic mutations, possibly caused by the effect of cosmic and ultraviolet irradiations. The theory claims that together, haphazard mutations plus natural selection are both capable over the eons to turn fishes into birds, and apes into Man.

At its core, the theory of evolution postulates that everything, from the start of life, till the emergence of conscious Mankind took place by purely physical means, with no supernatural interference whatsoever, that it was all by chance. Evolutionists also imply that the emergence of Mankind is not meant to be the *climax* of any alleged plan of progress or design. Mankind is a mere superfluous haphazard by-product of evolution, which has no significance or purpose.

Word by word, the creation story of the theory of evolution is as follows: Life was spontaneously created through random physical reactions. It has evolved to the present complexity through the power of blind, purposeless, natural selection acting on random-chance genetic mutations. Evolutionists have no difficulty with the process of natural selection per-se, as it has earned scientific proof. The only difficulty they have is to prove that random-chance genetic mutations can supply the optimum material for natural selection to accomplish its purpose.

THE KEYWORD IS RANDOM CHANCE.

> > > > >

CHAPTER 3

* * * * * * * * *

The History of Life
on Mother Earth

* * * * * * * * *

Figure (5), Mother Earth

HOW DO WE INVESTIGATE THE
HISTORY OF LIFE ON EARTH?

Leonardo Da Vinci (1452- 1519) was the first scientist to guess the importance of fossils in the study of life. He unearthed many fossils during the construction of water canals in Northern Italy. Three centuries had to pass before paleontology became a recognized branch of science. The basis for pale-ontological studies is fossils, which are remains, or traces, of organisms preserved in the Earth's crust. We find fossils everywhere, even in mountains which have been raised from the bottom of the sea millions of years ago. The two sciences of paleontology and geology got closely intertwined during the extensive research on how the surface of the Earth, and life on it, came to be.

Fossils were studied within the context of geologic time, and their relation to the rocks in which they were preserved. Radio-active isotope-dating is used for the assessment of the ages of the fossils and their geological strata, in terms of hundreds, thousands, millions or billions of years. When living organisms die they quickly decay into dust. Yet many remains can defy decay, such as bones, teeth and the calcified shells of land and marine animals. Teeth can tell lots of stories about the way an animal ate, its type of food, its habits and diseases. Bones tell its sex (male or female), how it walked, fought, and collected food. Whole intact animals or plants were found well-preserved buried under ice sheets, or in volcanic lava. Lots of intact insects were found fossilized in amber, which is a sticky secretion of the bark of certain trees. In some museums there are insects amber-preserved for 200 million years. They are hardly any different from contemporary ones even in the finest detail. Lots of pollen, spores and seeds, as well as microscopic forms of life have been documented, dated, and fully studied. In the last two centuries, paleontology benefited from the general improvement in methods of scientific research, especially computers and electron microscopes. Widely dispersed small data kept falling into place in a huge jigsaw puzzle. Paleontology became not only a study of ancient plants and animals per se, but also a study of the sequential development of all organisms through time. From the available data it has become scientifically possible to give the accurate chronology of not only the life on Earth, but also the story of the Earth itself.

THE STORY OF THE EARTH

From the vast wealth of scientific knowledge now available we know that our Earth was formed 4,500 million years ago. It was originally a part of a huge halo of interstellar dust clouds, as we will explain in the next chapter. The fragmentation of this dust cloud created the solar system: a Sun, around which ten planets keep on touring day and night. We now understand how this solar system materialized, as a part of a huge Universe. It has proved to be a masterly execution of a massive *'Master-Plan'*, initiated 15 billion years ago. This and the next chapter will give the details. It will also explain how our bodies and minds actually got *COOKED* up there, in the stars.

After formation, the Earth started to cool down for a billion years, but didn't freeze. This is because of the presence of radioactive isotopes with half-lives of one to ten billion years. They provide a continuous source of heat within the earth, in addition to the heat provided by the Sun. The Earth had an atmosphere full of water vapor, and various gases, but not oxygen. Water started to fill rivers and oceans in the form of rain. We now know that an important source of that rain was extraterrestrial. Tens of thousands of small icy comets, few meters in diameter each, fell onto the Earth every day. University of Iowa physicists estimate that this cosmic sprinkling added one centimeter of water level to the oceans every ten thousand years. Multiplied by the Earth's age of 4,500 million years that could account for all the water we have.

The diameter of the Earth is about 6,000 kilometers. The innermost core is made of iron solidified by pressure, covered by an outer core of molten hot iron and heavy metals. Outermost is the crust, about 20 kilometers thick, floating on a softer more malleable layer called the mantle. At first we thought that the continents had been in their present shape all the time, but in the 20th century we knew otherwise. The earth's outer shell is actually broken up into large rigid fragments called plates; they are partly submerged under the sea level and partly above water as land. The earth's crust is embedded within the plates, and the plates move over the fluid-like mantle. The study of how the continents took their present shape is called plate tectonics, or continental drift. The energy driving this massive movement of the continents is derived from the convection currents of the molten deep layers of the earth. Up to 250 million years ago, there was a super continent, now called Pangaea. It broke up into a northern part, that later formed North America, North

Europe and Asia, and a southern part, that later formed Antarctica, India, Australia, Africa and South America. Later on, the Indian subcontinent drifted northward to push against Asia and raise the Himalayan Mountains. Australia drifted northwards to its present site. Africa drifted northwards and pushed against Northern Europe, raising the Alps range of mountains, and forming the Mediterranean Sea. The Americas got connected, and now both North and South America are drifting together westward away from Europe and Africa. The Atlantic ocean is widening at the rate of one inch every year. The Pacific Ocean is shrinking at about the same rate, and ultimately America will push also into Asia. Antarctica is slowly rotating anticlockwise, around the South Pole.

When life started, no one was there to document the facts. As mentioned, the Earth materialized four and half billion years ago. One billion years later the oceans became full of single-celled living beings (blue-green-algae) which contained a substance called chlorophyll. This substance gathers solar energy and turns it into complicated carbon substances, as well as atmospheric oxygen. Three thousand million years later, there was enough oxygen in the atmosphere of the Earth for animals to be able to breathe. Then, *almost suddenly,* millions of more complex forms of plant and animal life started to appear, to diversify, to multiply, and to fill the Earth over the last short period of only 500 million years.

What the unicellular algae were doing for three billion years was just setting the stage for the more complex forms of life to appear on Earth. Their sudden appearance is sometimes termed 'The Big Bang' of evolution.

How did the blue-green-algae materialize at the very start, out of molecules of inert elements? The body of a dead organism contains exactly the same elements present in the living one. The only difference is the ability of the living one to further metabolize the elements of water, carbon, nitrogen, and oxygen; and, more importantly, to replicate itself. These are the two characteristics of what we call *LIFE.*

We will never ever know how the single-celled life materialized on our Earth. These cells are already quite complex as they are not just a collection of amino acids and chemical compounds. Anyone studying the universe will immediately notice that an insect, or even a single cell, is much more complex than a planet, a star, or even a whole galaxy. Many hypotheses were contemplated to explain the first appearance of single-celled life on our planet. The first suggestion was the action of electric energy from lightning strokes fusing some atoms together; until they form complex molecules...*which would ultimately be able to replicate themselves*. The last half sentence is formed of eight words. It is just playing around with words, because it gives no real explanation. After thousands of highly publicized experiments, and computer simulations, that speculation looked very odd indeed, and was gradually dropped. Another guess was the way snow flakes form. If you collect these flakes on a black cloth you can admire their shapes, and even photograph them before they start melting away. It was noticed that in every snow drizzle all the snow flakes look all alike in a certain shape. This observation was argued in favor of how molecules could replicate themselves. But soon the very idea was proved false. The shapes of *all* the falling snow flakes are pre-determined each time by the temperature and the degree of humidity in the sky, they are not replications. Finally, there was the suggestion that the seeds of life may have come down from outside the Earth. They could have reached here on board the thousands of small comets which hit the surface of the Earth every day. The idea is called *Panspermia*. But that was no answer at all. It even complicated the whole issue for our scientific way of thinking. It simply deferred the question to some extra-terrestrial scientific society (!!).

In brief, the blue-green-algae were followed one billion years later by bacteria. They co-existed with the algae in a dull single-celled monotony for a further two billion years. The monotony was suddenly broken up by what we call The Cambrian Explosion of complex multi-cellular forms of life, which started at about 540 million years ago. The first fossils from the time 543 to 490 million years ago were found in the Cambrian Hills in Wales, hence the name. Most of the presently known phyla of life classification became apparent within the span of five to ten million years. [Parker, 2004: Since the Cambrian no *new* animal phyla have ever evolved.] This *short* span of five to ten million years was almost instantaneous, in geologic terms of time. Creationists say that this suddenness was a miraculous act of God. Evolutionists squabbled much among themselves about several explanations. A much more feasible, *and scientific*, explanation is presented in Chapter 7.

After the Cambrian explosion the living phyla of multi-cellular life started to diversify. The first to appear were the marine invertebrates and marine plants, followed by snails and starfishes. The amphibians, vertebrates, fishes, insects and land plants came 100 million years later.

Forests and dinosaurs came about 180 million years ago, followed by birds and flowering plants. A massive extinction of many life forms, especially the dinosaurs, took place 65 million years ago. It is explained by the catastrophic impact of a meteorite hitting the Earth. A few million years later came the mammals, followed by all recent plants and animals. It is worthwhile to mention that almost 90 % of all multi-cellular species that ever existed have gone into extinction, within a few million years of existence. We could identify the factors involved in many cases, but in most cases the explanations got buried with them forever. Interestingly, the last of every species was almost the same as its beginning, with only a few *micro* - evolutionary changes, affected by natural selection. This is a very important point in the never-ending debate between evolutionists and creationists.

THE DYNASTY OF SINGLE-CELLED LIFE

Multi-cellular life started only about 500 million years ago. It was continuously changing, and most of its products are extinct. In a glaring contrast, the Earth is still full of single-celled organisms, scarcely different from their ancestors which lived 3,500 million years ago. From their perspective, the story of life on Earth is one of continuity and stability. Their way of life has not been threatened by any environmental changes, and they are superbly adapted to the way they live. *"To them, the complexity of multi-celled plant and animal life is simply a curious and Incidental byproduct of their success."* [Gribbin, 1993, page 57].

Further, in a book titled *"Full House"*, Stephen J. Gould argues that life has always been dominated by its bacterial mode, including the blue-green-algae, which he considers to be nothing but photosynthesizing bacteria. He also stresses the significance of the anaerobic bacteria discovered in the ocean depths. They live on nothing but sulphur, basalt rocks and water. Single-celled life is the dominant form of life on Earth, *and NOT Mankind*, he further argues.

THE SELFISH DNA

This is a commonly quoted fanciful idea, sponsored by some evolutionists, [Dawkins, 1989]. In summary, it claims that the only valuable form of life is the gene. Genes live as temporary *passengers* in the bodies of every creature. Multi-cellular creatures act only as *vehicles* for their sets of genes. Their only aim is to survive in order to eternally transmit their valuable genes to future generations. The genes in every living organism create the blue-print of that organism in ways that help it outwit all potential predators and prey, just as a *selfish* way for the genes themselves to survive. The idea of selfish DNA is too fanciful to be taken seriously as science, nor as a philosophical thought:

Genes are all made of Adenine, Thymine, Guanine, and Cytosine. Having achieved the ability to replicate, those *'selfish'* genes could have continued to bask in a paradise of blissful existence and replication into Eternity. They could have continued their selfish existence, either as such bare genes, or at the maximum in form of single-celled simple organisms, which has proved to a very stable way of life indeed. From the survival point of view it was surely a very unfortunate decision to manufacture complex multi-cellular *'vehicles'* which would fight each other as predator and prey, endangering the very survival of their valuable *'passenger'* genes. If two men across the Pacific Ocean would get mad enough to start an atomic hell, then all the genes on this Earth could possibly perish. From square one, life supposedly would have to start uphill all over again.

VIRUSES

Many known viruses are non-pathogenic, but some of them cause dangerous diseases such as smallpox, polio, rabies, influenza and AIDS. Each virus is a packet of DNA and RNA covered by a coat of protein. It cannot replicate autonomously like bacteria for example. In this respect it can be comparable to a seed, which is a 'dormant' form of life that cannot replicate un-aided. In order to multiply, a virus has to enter a living cell, and shed its protein coat, thus baring its DNA and RNA. Following that, it uses the cell's own replicating machinery to reproduce new viral DNA and RNA and protein coat. The newly created viral bits assemble to become more viruses which can leave the cell to infect others. Viruses do not fossilize, so it is difficult to define when they first appeared on Earth. Their DNA and RNA sometimes add genetic material to infected cells. This could be one cause of mutations which influence the course of evolution of life. In genetic research and engineering, scientists use viruses as one tool to introduce various genes into living cells.

THE THREE BLANKETS AND
BALANCES OF LIFE ON EARTH

The Earth is protected by three blankets; the first is its magnetic field which diverts the deadly solar winds away. Solar winds are composed of electrically charged protons and electrons flooding the space of the whole solar system. If they are allowed free access to the surface of the Earth it could be lethal to many forms of life. The second blanket is the Ozone layer. Ozone is tri-atomic oxygen (O 3), and is formed by the action of the ultraviolet wavelength on ordinary Di-atomic oxygen (O 2). In that process it filters out most of the harmful solar ultraviolet rays. The original atmosphere of the Earth contained no oxygen; it was created later by the hard-working chlorophyll molecule of plants. It took the primitive plants called *blue-green-algae* 3,000 million years to fill our atmosphere with its 20 % of oxygen. Only then could animal forms of life start to exist. The third blanket is composed of carbon dioxide and water vapor; it is the green-house blanket. It freely allows infra-red heat waves to pass from Sun to Earth, but partially blocks its way back to the atmosphere. In this way a large proportion of the warmth of the day is kept through the night. This is unlike other planets which suffer baking heat during daytime, and then turn freezing cold once the Sun sets.

Earth is the only celestial body we know which simultaneously contains water in all its three forms: solid, liquid and gas. Water is also the only element in Nature which *increases* in size on freezing. Hence we see ice floating on top of liquid water, because its density is less. This unique characteristic is a *given* law of Nature and without it ice would sink to the floor of the oceans, and never melt. Soon the whole earth would become a lifeless frozen place. The ice caps which adorn the two Earth poles allow liquid oceans and marine life to freely flow underneath. Their shining surfaces reflect back some of the solar heat reaching the Earth, thus creating a very delicate balance. This balance has been a factor in the starting and ending of many ice ages in the history of the Earth. Liquid water often seeps through rocks and mountains. When it freezes it increases in size, creating tremendous forces which crack even the hardest rocks. This has been one more factor in shaping the geology of the Earth over millions of years. Oceans contain 98 % of the water on Earth. The remaining meager 2 % are all that is present in rivers, clouds, and glacial ice. Ocean currents are created by the energy of the wind, by the rotation of the Earth, and by the temperature difference between Tropical and Polar Regions. There are currents from north to south, from east to west, and others vice versa. There are also currents from surface to bottom of the ocean, and vice versa. These currents help stabilize the overall global temperature of the Earth, and help in supplying oxygen to deep-water life forms.

Water rises upwards in any fine tube by another *given* physical force which we call capillary-action. This miracle-like arrangement allows plants to pump up water from roots to tree tops. It was estimated that a tree a hundred feet tall would pump up more than a hundred gallons of water every hour. It achieves this remarkable feat in absolute silence and dignity. Just imagine the noise created by an electrical pump doing the same.

The first balance that maintains life on Earth is the legendary gaseous balance between plant and animal life. Plants take in carbon dioxide and create oxygen as waste. Animal life does the opposite; it takes in oxygen and creates carbon dioxide as waste. The second balance is that between herbivorous and carnivorous animals on one side and that between viruses, bacteria and parasites and their hosts on the other side. The third balance is the famous food chain. In this chain, plants are the only hard-working member. Plants act as a clever cook, who picks nitrates and phosphates from the soil, carbon dioxide from the air, and energy from the Sun. It churns them up, and creates sugars and proteins. Animal forms of life act as parasites, they bully their way into the kitchen area of our clever cook, and help themselves to ready-made plant food. Not only that, they start eating each other as well.

In the oceans and marine life the same food chain holds true. Here plant life is represented by phyto-plankton. These are microscopic plants which contain chlorophyll. They multiply on the water surface as well as down to depths within the reach of sunlight. They get eaten by zoo-plankton, which are also microscopic animal forms of life. Both plant and animal plankton form the bottom of the marine food chain. They are food for small fish, which in their turn are food for bigger fish, and so on. Whales by the way are not carnivorous; they feed on plankton, tons of it every day. The whale gulps several gallons of seawater into its mouth, and then squeezes it out through a special nozzle, making the famous fountain-like stream of water. In that process it sifts all the plankton content of the water and keeps it inside its big mouth. Jonah, for sure, was not palatable for the great whale which swallowed him; it soon spit him safe on the shore. [Jonah's, or a similar story, is present in *all* religious scriptures].

In the end, death claims all forms of life. Plants die in two stages, first their leaves keep falling to the ground every winter for many years, but ultimately the plant itself also dies. Death has its own kitchen work to finalize the food chain cycle, insects, bacteria, and worms finish up every dead tissue and turn it back into nitrates and phosphates.

EARTHWORMS

Earthworms deserve a subtitle of their own. They are funny little creatures, merely a few centimeters long each. They are the best bait for fish angling, which is why they are sometimes called angling worms. Their food consists of decaying organisms. As they eat, however, earthworms also ingest large amounts of soil, sand, and tiny pebbles. They burrow into the soil deep down to two or three feet, taking with them these organic decaying materials, thus actually fertilizing the soil. They also loosen the soil, letting air in. Moreover they also bring deep soil into the surface. It was estimated that every year as much as 40 tons of soil per acre are brought to the surface, in areas where earthworms are abundant. Earthworms are also direct food for a large variety of birds and animals, and indirectly they also provide food for Mankind by their beneficial effect on the soil. It is very evident that what these worms do far exceeds the necessities of feeding and reproduction. Their soil-moving habits are their last tribute to Mother Earth. Earthworms guarantee that the nitrates and phosphates are finally returned to the exact spot where they came from: *at the root level of plants.* If this function was not naturally performed by these little worms, it would take Mankind billions and billions of man-hour work to till the soil that way.

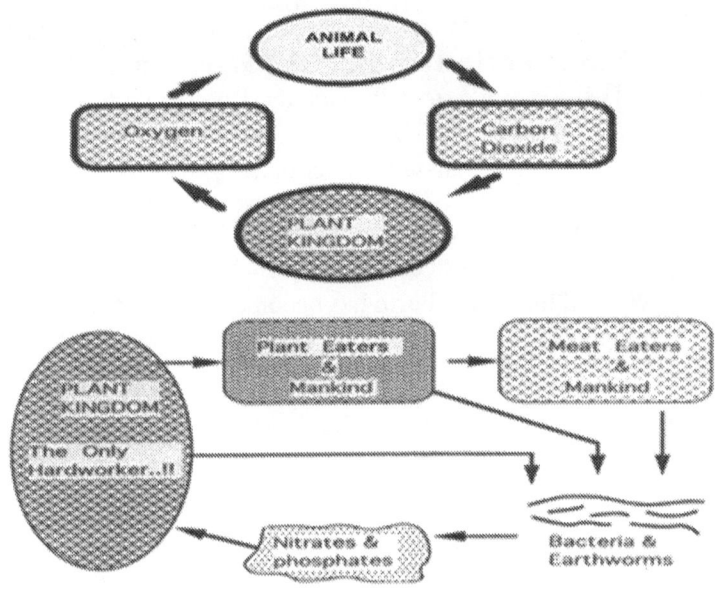

Fig. (6)
The whole food-chain is intricately inter-dependent. Take out any link of the chain, even the smallest, such as bacteria, plankton, or earthworms, and the whole cycle of life would grind down into a calamitous halt in a few generations. The Earth would turn into a barren, deserted and miserable planet, where no bird sings and no child cries.

SUMMARY OF THE RECORDED
HISTORY OF MANKIND

Five arbitrary and similar stages took place, *almost simultaneously* in various isolated human groups, on every continent around the world. Those stages are:

(1) The Paleolithic age, before 8,000 B.C.E. It was characterized by stone tools, and a life of hunting and gathering.

(2)The Neolithic age, from 8,000 to 3,000 B.C.E. Stone tools are still used, but farming is discovered, and people started to gather in towns and cities. They use written language, and constructed religious buildings such as ancestral statues, tombs, and temples.

(3) The Bronze Age, from 3,000 to 700 B.C.E. Bronze was discovered and metal tools made of it. Industry was added to agriculture.

(4) The Iron Age, from 700 B.C.E. to modern history up to the year 1,600 C.E. Discovery of iron and use of iron tools. Industry became well established. Historians like to subdivide the Iron Age into three distinct parts. Classical ages, from 500 B.C.E. to 500 C.E., middle ages, from 500 C.E. to 1500 C.E.; and Renaissance after that.

(5) The last stage is the one we live in. The main characteristic is the revolution in the means of communication, via land, sea, air, wireless, T.V., WWW, until the planet Earth has become a single huge village.

> > > > >

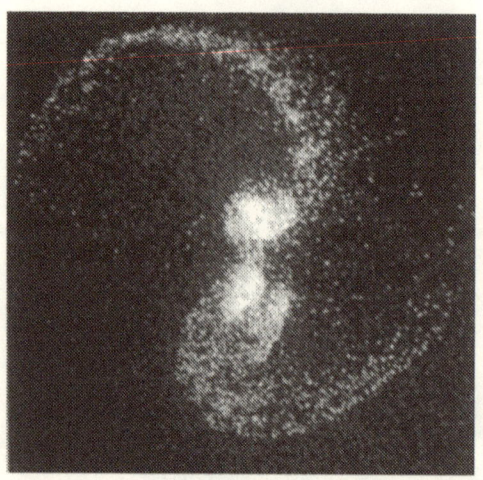

Figure (7)

Who would have thought that the huge size of the universe had any role to play in our own existence? In fact, for hundreds of years, philosophers have been using the vastness of the universe as an argument against the significance of life on Earth. Now, we know that on the contrary, the vastness of the universe is the basic essentiality for the creation of heavy elements, as well as all the necessities for creation of life on planets similar to our Earth. The only question that we may never be able to answer, is whether we are alone, or do we have company, somewhere?

CHAPTER 4

* * * * * * * * *

How We Got Cooked Up, In The Stars

* * * * * * * * *

All scientific knowledge started as mere "ideas" in the curious minds of people whom we later call "scientists". In his childhood, Einstein was intrigued by horses racing alongside a steam locomotive. In his youth he wondered what would happen to someone who is racing at almost the speed of light. Later on, in his adulthood, he was genius enough to answer his own question. It was the same question that led him to the "Theory of Relativity".

[After enjoying this chapter, you will surely feel much closer to GOD. But, there is a surprise for you in the last page, printed upside down, Don't read it now!]

Do you remember the blue party of the other day? (page 27), a few days later, we had a *black party*! It was a formal dinner on a yacht called Black Tiger, one mile off the beach. Men wore black ties, and the ladies had their pearls shining against black dresses. The beach lights were a mere faint line faraway. The stars in the sky also showed, glittering against a moonless pitch black sky. One lady was recently engaged, and was in a very romantic mood. Gazing at the stars, she wondered who could read her fortune. Another guest wondered if someone knew how far the nearest star is. The romantic lady volunteered the answer, it is four *light-years* away. We discovered that she had just passed a Master degree in Science, and that she also knew a lot about the sky. We enjoyed, very much, her soft voice telling us about the glittering diamonds up there, and the Earth down below.

WHAT IS THE UNIVERSE? AND HOW WAS THE EARTH   FORMED?

Once upon a time there were two apes asleep under an apple tree. One of them was a little cute, thickly furred, female chimpanzee. The other was a male naked ape, well dressed in a tuxedo coat and a top hat, his name was Isaac Newton. They were awakened by some apples falling off the tree. Miss Chimp grabbed two apples and ran away to enjoy a bonus meal. Sir Newton was reading a book; he put it down and started to ponder: why do apples always fall perpendicularly towards the earth? He started to scribble some notes on the back of his book, and in a few years time he announced his Theory of Gravity.

This anecdotal story took place in the year 1666. The theory had only two factors: the mass of any two objects, and the distance in between. It correctly explained thousands of observations made by the astronomy scientists of that time. It explained the layout of the Solar System, in which several planets eternally circle a bright Sun. It explained the lunar orbit and linked it with the ebbs and tides of the rivers and seas all over the Earth.

In the year 1915, the Great War was raging all over Europe, but the city of Berlin and its university were the safest place on Earth. Albert Einstein had no trouble at all conducting his research in the coziness of the Kaiser Institute. Since 1905 he pondered a badly needed modification to Newton's Theory of Gravity. At Christmas time 1915 he announced The

Theory of Relativity, which added a third important factor to Newton's theory, namely the factor of Time. Light was estimated to travel at the stable speed of 186,000 miles per second, which became the universal standard unit of time and space, but in 1915 there was no final proof. According to his theory, Einstein predicted that light can get deflected by the effect of gravity. During an eclipse of the Sun on December 12, 1919 The British Society of Astronomy confirmed this prediction, and this was the first proof of the most famous theory of the Universe.

A SIMPLE SKETCH OF THE UNIVERSE

Our Earth is one of ten planets that orbit around a star called the Sun. There are one hundred billion stars clumped together in a disc-shaped group called *The Milky Way* galaxy, our galaxy. The vast spaces of the Universe contain approximately one hundred billion more galaxies. Galaxies emit electromagnetic waves, which differ in their wave-length. The longest are called radio waves, next are the infrared (heat) waves, then the light waves, ultraviolet waves, x-rays, and finally the gamma rays which have the shortest wave-length. All such electromagnetic waves travel at the speed of light, which is 300, 000 kilometers per second, in a vacuum.

Our Earth is formed of the elements we know in the periodic table, starting from the simplest one, hydrogen, then helium, up to the heavier elements such as iron, potassium, mercury, bismuth, and radium, etc. Hydrogen has the simplest atom, consisting of one electron orbiting around a nucleus composed of one proton. An atom of helium consists of two electrons orbiting around a nucleus formed of two protons and two neutrons. Uranium has a nucleus formed of 92 protons and 146 neutrons.

Electrons are negatively charged, and their number in each atom is usually equal to the number of positively charged protons in its nucleus. Neutrons are not electrically charged, they are neutral. We would expect that the positively charged protons in each nucleus must repulse each other, breaking up the atom. But they are held together by a physical force called The-Strong-Nuclear-Force.

Neutrons are usually stable, but in radioactive elements they happen to decay. Each neutron then divides into three particles, and emits gamma irradiation at the same moment. The three particles are one proton and one electron plus a very minute particle called neutrino. The new proton is positively charged and stays in the nucleus. The new electron is negatively charged and escapes away at almost the speed of light, and is called beta irradiation. The neutrino is much smaller than the electron; it is neutrally charged and also escapes away at almost the speed of light. The physical force which breaks up radioactive elements in this way is called The-Weak-Nuclear-Force.

Recently, scientists discovered two important additional facts about atoms. The first fact is that even the small protons are formed of further smaller sub-atomic minutely sized particles. The second fact is that electrons sometimes behave as *particles*, and at other times behave as *electromagnetic waves*. This observation is the latest scientific dilemma and is called the Particle-Wave Duality. At *'the absolute zero',* (which is temperature minus 273 centigrade), electrons stop rotating around the nucleus, and all atomic activities come to a halt.

EINSTEIN'S THEORY

This theory states that Matter & Energy are two faces of the same coin; and are inter-changeable. This led to atomic bombs and atomic energy. It also postulates that the faster an object moves in space, the slower will time proceed for that particular object. If the object reaches the speed of light then time will stop for it altogether. You may happen to find it odd that time should slow for an object moving fast, but this has become an ordinary fact. Clocks on board of spaceships move definitely slower than those at the ground base. If the NASA scientists would ignore this factor in their computer calculations then a space craft would land several meters, or kilometers, off target. It would never be able to meet nose to nose with a Soviet space station and may get lost in space. If Miss Elizabeth Taylor would have volunteered for a space travel in the 1950s it would have been a great publicity for her latest film. If she travels at the speed of light for forty years then time would have stopped for her. She would land back in Hollywood in the year 1999 as young, beautiful and supple as she was in 1959, and would be ready to start a new theater career.

The Theory of Relativity gave much food for thought to authors who write science fiction, (for example, The Time-Machine). It also became the basis for hundreds of scientific research all over the world. When we look into the sky what we actually see is the past, not the present. We see the Moon as it was about one second ago, and the Sun as it was eight minutes ago. Our Solar System is just a few grains in a vast sea of sand called the Milky Way Galaxy. It is disc-shaped and is formed by one hundred billion stars, of which only seven thousand are visible by the naked eye. The nearest star is called Proxima Centauri, and it is four light years away. The image which comes into our telescopes today left that star four years ago. A light year distance is 9,460 billion kilometers. The farthest stars at the edge of our Galaxy are seen by us as they were one hundred thousand years ago.   There are millions and millions of other galaxies, exactly similar to ours. The only one visible by the naked eye is the Andromeda Galaxy which is two million light years away. It is double the size of our Galaxy, containing two hundred billion stars. The nearest Galaxy to ours is called the Large Magellanic Cloud, and is one hundred and eighty thousand light years away, but is not visible by naked eye because of its smaller size. One of its stars exploded into a Supernova and was visible to us for a few weeks time in February 1987. It gave astronomers everywhere a unique chance to test and prove many facts, as will soon be described.

HOW DO WE INVESTIGATE THE UNIVERSE?

Astronomers use many scientific tools to examine that vast sea of huge cosmic objects. One tool is the spectroscopic examination of the light rays emitted by each of them. This can define the chemical substances of which each of them is formed. While some are formed only of hydrogen and helium gas, others show the presence as well of carbon and oxygen, while others include heavy elements too.

In the year 1842 Professor Christian Johan Doppler of the University of Vienna described his famous observation which became known as the Doppler Effect. This theory connects the frequency of any type of wave with the relative motion between the source of the wave and its observer. When a star is rotating *towards* us its light rays will become sort of compressed, or shifted towards the blue part of the light spectrum, which has got a shorter wavelength. When it starts the other half of its orbit, rotating *away* from us then its light rays will become oppositely shifted towards the red part of the spectrum, which has got a longer wavelength. The Doppler Effect holds true not only for light waves, but also for any electromagnetic wave as well as for sound waves. If you listen to the whistle of an approaching train you will hear a higher pitch of sound. This will get muffled into a lower pitch when the train passes you and moves away. The red-shift of light from faraway galaxies also became an important measure of their distances, as will soon be explained.

Light waves are not the only connection between us and the rest of the Universe. Sound can never reach us because it travels only in gas or water, but there are lots of electromagnetic waves pouring into our scientific detectors from all parts of the Universe. Some are very similar to ordinary radio-waves, having wave-lengths from one meter to thousands of meters. Others are of much shorter wave lengths, e.g. microwaves, infrared waves, visible light, ultraviolet waves, X-rays, and Gamma rays, mentioned in their order of wavelengths. All these rays are detected by highly sensitive sensors, which can define their strength, direction, source, as well as any Doppler effect, if present. This has added a second scientific discipline called radio-astronomy to the previously popular one, called optical astronomy.

Using optical astronomy and parallax geometry, we were able to measure lots of distances in the Universe. For near objects we can use two different spots on the Earth as the base of the triangle. For further away objects we can use the whole diameter of the Earth. For even further spots we can use the distance between the Moon and the Earth, or the Sun and the Earth, as the base of the triangle. With the recent launch of space satellites and powerful telescopes, we can use the same principle for much more distant measurements in vast space.

Gravitational-wave experiments, as well as underground detectors for the particles called neutrinos have also added many valuable facts. Space research has also given a totally new dimension to astronomical observations. Space crafts can see the Universe in a much better way, unhindered by any weather changes or any obstacles in the atmosphere of the Earth. The human visits to the surface of the Moon have added positive physical samples which can be examined in a lab.

In summary, we get our knowledge of the Universe by fitting together millions and millions of small pieces of observations and scientific facts. The light and electromagnetic waves which bring this information have been initiated at various times in the long history of the Universe. Some are ten billion years old; others are just a few years or days young. Recently we have even been able to detect the extremely faint heat waves lingering in the cosmic background since their initiation at the beginning of the Universe, fifteen billion years ago.

OUR PRESENT KNOWLEDGE OF THE UNIVERSE

All these investigations and scientific developments have given Mankind a unique opportunity to behold an almost continuous strip of videotape which faithfully registers the way this Universe has come to be. This tape extends from billions of years back in our time to one second ago. Lots of very interesting facts have been registered on that tape.

In 1929 an American astronomer called Edwin Hubble gave our hypothetical videotape a remarkably dramatic start. He cleverly made a landmark observation: Wherever you look, you will notice that distant galaxies are moving rapidly, *away* from us, and also away from each other. Many billions of cubic miles of new space get included between the galaxies every day. Hubble's discovery went further. Not only are the galaxies moving away, but the speed at which they are doing so is getting greater the further away they reach. In other words the Universe is *expanding* in an ever *accelerating* way. With such expansion, the Doppler Effect will get bigger for more distant galaxies as they accelerate away. Their red shift will proportionately increase and will become a reliable measure of distance between those galaxies and our telescopes.

An expanding Universe means that at earlier times all of its objects would have been closer together. Scientific calculations showed that there was a time, about fifteen billion years ago, when they were all at exactly a single spot. At that spot, and at that moment, the Universe was infinitesimally dense, and its size was infinitesimally small, (a pinpoint singularity). That starting point of the Universe was called the Big Bang.

In less than one second of our time, that infinitely dense singularity exploded into a fireball with a radius of ten billion miles. That fireball was composed of nothing but sub-atomic-sized particles, which we now know as protons, neutrons, electrons and neutrinos. Today we know that the first three of these particles are the building blocks for atoms. In one minute, of our time, that infinitesimally hot fireball became like a giant thermo-nuclear reactor, transforming that endless array of particles into atoms of hydrogen and helium gas. An atom of hydrogen contains one proton and one electron, helium atoms consist of two protons, two neutrons and two electrons each. But they were not made in equal amounts, three quarters was made hydrogen, with only one fourth made helium.

Neutrinos are the fourth type of particles initiated at the time of the Big Bang. Recent research estimates that each neutrino has got the tiny mass of 5: 100,000 of the mass of an electron. This almost mass-less existence allows them to travel at almost the speed of light. They do not share in the formation of atoms; hence they do not share in the formation of galaxies, stars or the rest of the *visible* Universe. They are part of the 'Dark Matter', which constitutes more than 90 % of the calculated mass of the Universe. Recent research is further subdividing both the visible and dark matters of the universe into even more tiny particles with fancy names such as quarks, positrons, axions, photinos, wimps, mu and tau neutrinos, etc.

THE DARK MATTER AND THE DARK ENERGY

The latest scientific evidence suggests that only about five percent of our universe is composed of ordinary atoms, (protons, neutrons and electrons), that interact with light and hence are visible to us. This is referred to as the Baryonic part of the universe. Even later in the history of the universe, some of this visible five percent lost their luminosity and became detectable only by their gravitational effect. Examples are the white and black dwarfs, neutron stars and Black Holes, soon to be described.

Now, where is the remaining 95 % of the calculated mass of the universe? One fourth is dark matter such as neutrinos that do not interact with light, neither absorbing nor emitting light. Neutrinos are not detected by vision, but by their gravitational effect (attraction) on the visible objects of the universe. At the moment of the Big Bang neutrinos acquired a tremendous momentum, which sent them at almost the speed of light in all directions, expanding the Universe and *dragging* with them the much heavier giant cosmic cloud made of hydrogen and helium gas.

Seventy percent of the measurable mass of the Universe is still missing. However, in the last five years, observations have shown that seventy percent of the bulk of the universe is in the form of a ubiquitous dark energy with a strange but remarkable feature: *Its gravity does not attract, it repels.* Where gravity pulls the chemical elements and dark matter into stars and galaxies, it pushes this dark energy into a nearly uniform haze that permeates all space. This *repulsive* gravity is the main factor in the accelerating inflation of the universe. Einstein felt its repulsive effect in his calculations, but he didn't understand it at the time. The fact of accelerating inflation of the universe was only identified by the Hubble telescope two decades later.

Before the fateful minute which followed the Big Bang we have no way of knowing or even pondering, about matter, time, space, or laws of physics. But from that minute on until today, everything has behaved in a way which faithfully obeyed all the laws of physics, electromagnetic, and mathematics which we now know. This is a remarkable bit of scientific knowledge. It simply means that the whole Universe had a definite *beginning* in both space and time. It also means that every tiny particle and every object in this Universe does obey certain *Laws of Physics* that have been validated at that initial moment of the Big Bang. Laws of physics are simple enough to be comprehended by the human mind. The rules which we deduce from experiments here on Earth seem to apply on every object in the whole Universe, not only today but since Time actually began.

This remarkable bit of scientific knowledge smacks very much of being the action of a willful and intelligent force. At a certain moment that force created the particles out of which matter, and life, will later form, and at the same time gave those particles the rules with which they will play the game.

WHERE DID WE COME FROM?

This is the ultimate question. If that ten-billion-miles ball of fire was the actual nucleus of this whole Universe, and if it was originally formed only of a tremendously hot cloud of hydrogen and helium gas, where then did the elements forming our bodies come from?

Repeated computer model studies of the matter of the Universe, and the physical laws starting from the second minute after the Big Bang, have shown the impossibility of making anything other than hydrogen and helium. The only physical force acting at that moment was the tremendous heat *initiated* at the Big Bang. Its traces are still faintly detectable today, lingering in the cosmic background, fifteen billion years later. The only atomic particles available then were protons, neutrons, electrons and neutrinos. The only thing that heat can do, without the addition of other forces such as gravity or magnetism, is to unite a proton and an electron to make a hydrogen atom, and unite two protons, two neutrons and two electrons to make a helium atom. These facts coincide with every physical and astronomical observation made until today. The fact that we exist shows that carbon, oxygen, nitrogen and the other elements have been manufactured somewhere, at later times, and then dispersed through Space in order to share in the formation of our home, the Earth. The only place in which heavier elements could have been manufactured, and are still manufactured till today, is inside the stars.

> > > > >

But what are the stars, and how do they work?

THE KITCHEN-WORK IN THE STARS

By this term we mean the mechanism by which the various elements get manufactured inside the stars. The later development of the Universe passed through various stages which could be identified as infancy, adolescence and adulthood. During the infancy stage of the Universe it was a giant expanding cloud of cosmic dust. The cloud consisted of 76% hydrogen and 24% helium gas as well as the Dark Matter of neutrinos. The neutrinos traveled at almost the speed of light, continuously expanding the size of the baby Universe, and dragging with them the much heavier hydrogen and helium gas. After a few hundred thousand years, (of our time), the smooth distribution of the cloud started to fragment into large groups which became the seeds of later galaxies. Recently the orbiting Hubble space-telescope estimated the total number of galaxies at one hundred billion, which is five times greater than we believed previously.

Within each galaxy the gas cloud further fragmented into smaller pieces, the gas in each of which started to clump together under its own gravity to form an embryo star. These are called first-generation stars. In about fifty thousand years a dense very hot core has formed for each star under the effect of its own gravity. The heat in the core started nuclear fusion of hydrogen nuclei to helium nuclei, with release of some energy from each single fusion. The basic source of energy for all stars is the same. It comes from nuclear fusion which builds heavier atoms from lighter ones. The first step is turning hydrogen nuclei into helium. When the outward pressure of the energy of this hydrogen furnace balances the inward pull of gravity of the total mass, the star is said to have become *mature.*

There are thousands of millions of such mature first-generation stars in every galaxy. Each star maintains this balanced status quo for ten to twenty million years until all the hydrogen in its core has been exhausted. After that, any one of many alternatives may take place. Every one of these alternatives leads to a certain cosmic occurrence which has been repeatedly observed, documented, and explained. Which alternative occurs depends on the original mass of the star. Our Sun is a second generation star, rather different from what we are talking about now. But for the sake of simplicity astronomers use its size and mass as a reference unit of measure called *solar mass.* The radius of the Sun is 430,000 miles, (109 times the radius of the Earth's 3,950 miles). Its total mass is 330,000 times the mass of our Earth.

THE SOURCE OF ENERGY IN A STAR

In 1983 the Nobel Prize for physics was awarded to William Fowler, an American nuclear physicist. He worked closely with Sir Fred Hoyle of Cambridge University of Britain in an extensive study which cleared the secrets about the formation of heavier elements. The trick is simply sticking helium nuclei together, *against the strong electrical repulsion of their positively charged protons.* Heat is not enough to do that, the force of gravity is called upon to act as a pressure-cooker, and this can take place only in the more massive stars.

The nucleus of Helium-4 is a stable particle, sometimes called Alpha particles. It is formed of two protons and two neutrons. When two of these Helium-4 particles are forced together you will get a Berryllium-8 nucleus. If a further one is added we will get a Carbon-12 nucleus. Carbon is the most crucial element for all forms of carbon-based life like ours. Adding a fourth Helium-4 nucleus will give an Oxygen-16 nucleus. Oxygen is the life-line we breathe. Fusing two Carbon-12 nuclei together will produce a Magnesium-24 nucleus. Adding a further Helium nucleus will give a Silicon-28 nucleus. Fusing two Silicon-28 nuclei together will give the very stable Iron-56 nucleus.

The kitchen work inside any star is just as simple as that, provided the mass of the star is big enough to exert the amount of gravity-pressure needed for every step. A star of one to three solar masses will usually stop at the level of Carbon-12 production. A star with double that size may reach up to Silicon-28. Another with still bigger size can end up as a giant crystal of Iron-56 floating in space, with nothing more important taking place. Such inactive stars lose their luminosity, they are called black dwarfs. They simply join the host of Dark Matter in the Universe.

THE DEATH OF A STAR

However, when the original size of the star is 20 to 25, or more, solar masses, then something completely different takes place. It is a very exciting scenario, which can be summarized in clearly defined successive steps. The star will keep burning hydrogen into helium, in a balanced mature state, for some tens of millions of years. When hydrogen in the core is exhausted, the outward pressure of the burning energy will diminish. Gravity will collapse the star on itself to a smaller size, and will exert enough pressure to start fusion of Helium-4 nuclei into Carbon-12. This will continue for a few million years until the helium is exhausted. The amount of energy released by helium fusion is much less than that of hydrogen fusion. The star will further collapse, creating more pressure which forces Carbon-12 nuclei to stick together. The energy released is much less, and most of the carbon is consumed in no more than 600 years. Next to occur is Neon-20 fusion, which may end in just one year's time. Oxygen-16 fusion will follow, with less and less energy released, and will come to an end in just six months. The star is now greatly collapsed under its own gravity. With each one of these steps the core is getting hotter and hotter, and the nuclear fusion reactions are becoming more and more violent. The greatest reaction is Silicon-28 fusion into Iron-56. This occurs in a flash of only 24 hours, after which a *supernova explosion* takes place.

This is a remarkable cosmic incident, which takes place approximately every 400 years in our Milky Way galaxy and at more or less the same frequency in other galaxies too. A memorable Supernova took place in the year 1054 A.D. when a star exploded in the Crab Nebula. This Nebula is only 4,000 light years away, and is a part of the Taurus Constellation in the northern hemisphere of the sky. For a few weeks that star shined almost like a little Sun and was visible with the naked eye. This extraordinary celestial event was recorded in human history long before it could be scientifically explained, *and has given rise to many mythological stories and legends.* No one at that time knew that the light they were beholding was coming from an accident which took place 4,000 years earlier.

The most recent and well-studied Supernova explosion took place only a few years ago, in February 1987. The exploding star was located in a nearby galaxy called the Large Magellan Cloud. The star was named 1987-A, and was 170,000 light years away. For a few weeks it shined into our telescopes with the luminosity of a whole galaxy. It gave astronomers and scientists a unique chance to confirm and elaborate on many aspects of this remarkable celestial event. The studies showed how the final stage of Iron-56 suddenly turns into a calamity which starts and ends in less than a second of our time. Under the tremendous pressure of gravity, the newly formed iron core of the star suddenly breaks down. All of its protons and electrons are forced to merge together to form neutrons. The whole core of the star suddenly gets reduced to a much smaller sized lump of neutrons. In spite of the tremendously reduced size, the mass of that lump of neutrons is still the same, hence its extreme density. At this point, the outer layers of the star suddenly fall in to fill the void. Their mass, which is almost equal to about 24 Suns, suddenly converges at a speed almost 15% of the speed of light to bang against the tiny in-compressible lump of neutrons. This unimaginable clash causes an enormous heat *and* shock wave which bounces back to explode the outer layers of the star. The pressure of that tremendous shock wave far exceeds the gravity pressure of the whole star. It cooks the helium in the exploding outer layers into carbon, oxygen, nitrogen, sodium, chlorine, sulphur, calcium, potassium, etc. It does not stop at the level of Iron-56, but proceeds to heavier elements such as gold, lead, silver and uranium, and all the elements in the periodic table. The nuclei of these heavier elements are heavier than Iron-56, but are less tightly packed. To make them out of iron *more* energy has to be spent, rather than released. This is why the pressure-cooker in the core of any star always simply runs out of fuel when it reaches the stage of Iron-56, inviting the collapse of the star.

PULSARS ...THE LITTLE GREEN MEN FROM SPACE

With such an accident the star gets reduced to a tiny core. This core has a mass equal to almost one solar mass, tightly packed into a size no more than a few miles in diameter. It is formed of a densely packed lump of neutrons, and starts spinning around itself at rates up to 200 times per second. It keeps on doing that forever, emitting strong radio waves from its magnetic poles. These waves sweep through space like the beams from a lighthouse, and are detectable by radio telescopes as regular pulses, earning the name of Pulsar, or Neutron Star. There are millions and millions of Pulsars throughout the Universe. Each of them is a memorial of a previous Supernova explosion. Each of them ticks its pulses with very precise timing. So much so that astronomers now use them as cosmic clocks, even more accurate than the atomic clocks which we artificially make.

These pulsars were first observed in 1967 by Jocelyn Bell, a research student at Cambridge University. She convinced her supervisor, Anthony Hewis, that she had just made a first contact with an alien civilization in our Galaxy. By more careful observation they located four of them, and they presented that discovery to the Cambridge astronomy seminar under the exciting names of Little Green Men No. 1, 2, 3, and 4. The news made a strong impact in the media, and soon Hollywood's science fiction writers were contemplating a first Space-Western film. In less than one year's time the much less romantic facts were elucidated, and the neutron stars were given their present name of 'Pulsars', but Miss Bell's first four did retain their additional honorary names, L.G.M. No 1, 2, 3, and 4, on the map of the celestial hemispheres.

THE FORMATION OF HEAVY ELEMENTS

The remainder of the exploding star, about 24 solar masses or more, becomes a giant cloud of gas expanding into space. It is called a second-generation cloud, and is more or less similar to the first-generation gas cloud originally formed at the time of the Big Bang. This time the gas is formed not only of hydrogen and helium, but also of all the elements known to us on Earth. These elements represent only two percent of the mass of the expanding cloud. Helium represents its usual 24%, and the remaining is hydrogen. The expansion of this gas cloud into space is very reminiscent of the expansion of the original gas cloud initiated at the Big Bang. Similarly also it cannot go far, unless *dragged* over by neutrinos traveling at almost the speed of light.

It was to the delight of all astronomers studying the Supernova explosion of the year 1987 to discover that their calculations and predictions were correct. The catastrophic collapse of 24 solar masses squeezing the tiny neutron core actually causes the emission of a flood of neutrinos, a fact unequivocally proved by several observatories on the Earth in February 1987. The flood of tiny-massed particles, traveling at almost the speed of light, guarantees the total explosion of the outer shell of the Supernova star. It also carries the resulting gas cloud into the far reaches of Space.

The second-generation cloud, same as the first-generation clouds of the Big Bang, will start to aggregate into clumps. These clumps form embryo *Second-generation* stars like our Sun. This time the newly formed star will have everything it needs: hydrogen fuel to turn on its own light, as well as a halo of gas rich in all the elements. This halo will cool down and surround the new star with several planets; one of these planets was our Earth.

"Who would have thought that the huge size of the Universe had any role to play in our own existence? In fact, for hundreds of years, philosophers have been using the vastness of the Universe as an argument against the significance of life on Earth." [Barrow, 1995]

Now, we know that on the contrary, the vastness of the Universe is the basic essentiality for the creation of heavy elements, as well as all the necessities for creation of life on planets similar to our Earth. The only question that we may never be able to answer, is whether we are alone, or do we have company, somewhere.

OUR SOLAR SYSTEM

The mass of the Sun is 333,000 times that of the Earth. It actually represents 99.9 % of the mass of the cloud which has formed the whole solar system; the remaining 0.1 % is shared by all the ten planets. Four of them, Mercury, Venus, Earth and Mars got a much bigger share of elements heavier than hydrogen and helium; hence they are called terrestrial planets. The rest are called the Jovian planets, Jupiter, Saturn, Uranus, Neptune and Pluto, and they are much more gaseous than the first four.

The distance between the Sun and Earth is about 93 million miles and is used by astronomers as an astronomical unit of distance. The light year distance is 9,460 billion miles, and is used as a more suitable unit of distance measurement beyond our immediate neighborhood in the Universe. The Sun is 4.6 billion years old, it is in mid-age now. The temperature in its core reaches 15 million degrees centigrade, and the density of the compressed gas is twelve times that of lead. Inside that core millions of tons of hydrogen are fused into helium in every second, a mass equal to five million tons of matter is converted into energy and dispersed into space to light and warm our Earth.

Four billion years from now the hydrogen fuel in its core will be exhausted and hydrogen burning will spread into the outer layers of the Sun. It will then expand into a vast red giant engulfing the inner planets including Earth. All forms of life on the Earth will surely have perished in a further billion years, and our Sun will have become a little dwarf hardly giving any light, and will disappear from the *visible* sky.

Spectroscopic study of the Sun shows that it is composed of 72% hydrogen, 26% of helium, and two percent of all the heavier elements we know, including iron and uranium. One of the elements which formed our Earth is radioactive uranium. This element decays at specific well known rates.   Examination of Uranium cores on the Earth has defined its age at between 6 to 15 billion years, which corresponds to the formation of Supernova explosions in the early age of the universe.

Study of the cosmic radiation which bombards the Earth day and night shows that there are two types of radiation. The first is the electromagnetic radiation, such as radio waves, x-rays and gamma rays. The second type is particulate radiation which is composed of nuclear particles. These particles are predominantly, (87%), protons, which are the nuclei of hydrogen and 12% alpha particles, which are the nuclei of helium. There are also nuclei of lithium, beryllium, carbon, nitrogen, iron, etc., but at fifty times greater abundance than their abundance in the inter-galaxy spaces. Their source is believed to be from repeated Supernova explosions in our own Milky Way galaxy.

ARE THE HOLES REALLY BLACK?

If the collapsing star in the above scenario is much larger than fifty or one hundred solar masses, then the end result will be a much different story. Gravity takes the upper hand, collapsing the whole matter and energy of the star not into a neutron star, but further, into a *singularity*, a mere single spot in the vast void of space. Space-time becomes bent, and escape velocity will exceed the speed of light. The escape velocity of any star, or planet, is the speed needed for any object in order to be able to overcome the star's force of gravity and escape into outer space. The star will simply get *shut off* from the rest of the universe. Not even light can escape its tremendous gravity force. It becomes a Black Hole, termed so by physicists Roger Penrose and Stephen Hawking.

The Black Hole starts sucking up loose gas and dust from nearby galaxies. The sucked gas and dust frantically emit X-ray waves just before plunging into the oblivion of the Black Hole. A funny name was given to this phenomenon; Black Holes were nicknamed the *'vacuum cleaners'* of the universe, vacuum cleaners whose bags never need changing. Black Holes abound everywhere in the universe. Their calculated tremendous mass and gravity are deduced from their gravitational effects on the rest of the visible universe. It is calculated that there is a huge black hole at the center of our Milky Way galaxy, around which the galaxy rotates. Its mass is calculated to be about one million times the mass of our Sun.

[In every book about cosmology, (academic or popular), you will always find a few pages describing what would happen to an observer in a space craft drifting towards a black hole. The story gives a lot of thought to writers of science fiction. His radio messages received at the NASA headquarters would become gradually muffled then disappear altogether. Radio waves, same as light waves cannot escape the tremendous gravity of the black hole. The observer himself would hardly notice that anything weird is happening at all. For him, Time will become slower until it completely halts when he reaches the speed of light. In the flash of a second his body and craft will become shredded, not into pieces but into atoms, **then into sub-atomic particles**, which would then drown into the singularity of the black hole. It will be as if the poor man and his craft have got thrown into a trash bin *outside* the universe. In the last momentous second of his life he will have, first handedly, witnessed a live video-film of the Big Bang itself, with the video running *backwards*!! Unfortunately, he would carry that secret with him into the oblivion of no return.]

CHAOS AND THE QUANTUM-UNCERTAINTY PRINCIPLE

The word 'Atom' is derived from two Greek syllables a - tom which means the thing that cannot be divided anymore. Modern science is reaching now to the study of *sub-atomic* particles. We now know that the neutrons and protons are formed of even smaller particles. We also know that electrons do sometimes act as *particles*, and other times as *waves*. If you run them through an apparatus designed to see particles, you see particles. If you run them through an apparatus designed to see waves, you see waves. This is the wave-particle duality, which is the latest dilemma of physics. We are approaching the proof that matter and energy are actually two faces of the *same* coin. In experiments at this sub-atomic scale we use equipment that either measure electromagnetic waves, or measure the position or momentum of particles. In itself the act of measurement interferes with an existing wave or an existing particle. Hence, this is one of the difficulties of this discipline, in the world of science.

Physicists are now looking for the 'Theory of Everything' that would unite the four known physical forces, namely Gravity, Electromagnetism, Strong and Weak Nuclear Forces. In their search, they are theorizing that all sub-atomic particles are made of vibrating 'Strings'. The difference between one particle and the other is the 'resonance' of its string. There is no feasible way for experimental work to prove, or disprove, this so called 'String Theory'.

In his book *Nature's Numbers,* page 125 Ian Stewart says that radioactive atoms decay at random, their only regularities are statistical. A large quantity of radioactive atoms has a well-defined half life, (which means the period during which half the atoms will decay). But we can't predict **which** half. By the way, Einstein's protest about God playing dice was aimed at just this question. There must be a difference between a radioactive atom that is not going to decay, and another which is just about to. *Otherwise how does the atom know what to do?* The statistical regularity is the trace of the underlying determinism. We just don't know which atom will decay because of further **hidden variables** not known to us. Where else can statistical regularity come from? [Stewart, 1995]

A butterfly in Tokyo flapping its wings may result in a hurricane in Florida one month later. This has become a common expression in modern science under the term 'Butterfly effect'. This is not chaos; it is just hidden factors not fully known to us, yet! The route from cause to effect becomes so complicated that no one can follow every step of it.

Nobody suggests that *statistics* are *unscientific* because they deal with unpredictable events. Chaos should be treated in the same manner. Nature's chaos is governed by rules, even if they do not show patterns that are obvious to us. There are simple laws right under our noses, laws governing everything from disease epidemics, to heart attacks, to plagues of locusts. If we could learn these laws, we may be able to prevent some human disasters and even disease.

WHAT IF?

All the scenarios described so far have ended in the formation of planets such as Earth, hospitable enough for Carbon-based life like our own to develop, and for men and women to start asking questions. *This would have never taken place if ever there was the slightest change in any of the particles initiated at the Big Bang*, whether in their masses, properties, or the physical laws which they obeyed. Computer models and calculations of high sophistication have repeatedly shown a number of very interesting scenarios.

A completely different universe than the one we have would have developed if the force of gravity was just a little stronger or a little weaker. A totally uninhabitable universe would have emerged if the electric charge of the electrons, or protons, were a tiny fraction different. Similarly, our universe would not be the same if the temperature during the fateful minute following the Big Bang was a few degrees less or a few degrees more. Were gravity just a bit stronger, cosmic expansion would have halted, and the universe would have collapsed before life could have evolved anywhere. The sun would have lasted only for a mere billion years, and planets would have lit up and become mini-stars. If, on the other hand, were gravity weaker, then no galaxies, stars or planets, would have had the chance to form out of the original cosmic cloud. If protons were just 1% heavier they would spontaneously decay into neutrons. In this case, no hydrogen atom could exist and no stars could light up.

Carbon atoms are the basis of life on Earth. They are made inside the stars by the union of three helium nuclei. This feat critically depends on what scientists call 'the internal resonance' of carbon and oxygen nuclei. Oxygen is formed by the union of four helium nuclei. If the carbon resonance was only a few percent lower, or the oxygen resonance only 0.5 percent higher, then no carbon would have ever been formed.

THE ANTHROPOMORPHIC UNIVERSE

In their book *"The Stuff of the Universe"* John Gribbin and Martin Rees say: "The basic physical laws have been *given* as such, and have never been modified. They did not *evolve* like the evolution of living organisms in tune with their environmental change." In the final conclusion for their book is the wonderful statement that "it looks as if the Universe has been *tailor-made* for Man". Another popular quotation by physicist Freeman Dyson says that in some sense it looks that *"The universe knew we were coming"*. The idea earned itself a special term, which was coined as the 'Anthropomorphic Universe', which means a universe which is *friendly* to Mankind. One of the best known of Einstein's sayings is quoted as such: "The most incomprehensible thing about the Universe is that it is comprehensible to us. The basic physical laws which our brains can understand seem to apply not only on the Earth, but also everywhere."

Stephen Hawking, who now holds Newton's Chair as Lucasian Professor of Mathematics at Cambridge University, is widely regarded as the most brilliant physicist since Albert Einstein. He is quoted to say: "No set of equations can explain why there is a Universe. We may understand the equations and the physical laws, but will never understand the mysterious Force which is breathing fire into them."

In his popular book, *"A Brief History of Time"*, Professor Hawking says that so long as the Universe had a beginning, we could suppose it had a Creator. He also says that the initial state of the Universe, at the time of the Big Bang, must have been very carefully chosen indeed. It would be very difficult to explain why the Universe should have begun in just this way, except as the act of God, who intended to create Beings like us.

The whole history of science has been the gradual realization that events do not happen in an arbitrary manner, but they reflect a certain underlying order, an order which is very intelligent indeed. The equations and physical laws are comparable to the set of rules of chess. It takes a Master's mind to breathe fire into those rules, and play and win a game.

The facts described so far are well known to everybody, including atheists. I believe that their response is very interesting. The term 'atheists' refers to several groups of people. First, there are scientists who sincerely believe atheism in their hearts. Second, are communists and ideologues who *need* the idea of atheism, because 'Theism' would be an absolute obstacle against what they want to do. Third, are the silent majority of people of all societies and creeds, who are baffled by the quasi-scientific jargon of the first two groups. This silent majority gets very little help from the teachers of Theism of any creed in the world. The response of those teachers is commonly in the form of enforced dogmatic interpretations of their specific Scriptures, for which any argument is taboo. To this silent majority, I feel the deepest sympathy, and will try my best to offer them a much better option. That is the sure knowledge that the deeper we go into scientific facts, the more we realize the magnitude of the intelligent harmony behind all that. This is a harmony which bespeaks only one fact, the fact of an external intelligent force, which has initiated it all.

We have just summarized numerous scientific facts about our own universe, from its beginning, through infancy, adolescence, until its present adulthood. The first response of the atheist camp was denial. The discovery of the fact of the Big Bang gave the Universe a point of beginning, which smacked very much of a Creator's act. They rushed to deny and discredit any fact pointing to the Big Bang. However, they did not present any new fact, instead they started a campaign of theories which had only the aim of distracting scientists and lay people.

In his book *"Brief History of Time"* page 47 Stephen Hawking says that many people did not like the idea that Time had a beginning. There were therefore a number of attempts to avoid the conclusion that there has been a Big Bang. The proposal that gained the widest support was called 'The Steady State Theory'. Another theory of Atheistic denial is described by John Gribbin and Martin Rees in their book *'The Stuff of the Universe'*. Its chapter eleven is titled 'Off the Peg'; it talks about a suggestion that there are many Universes, not only one.

May I take this chance to stress here one of my strongest beliefs. I believe that at least some of these attempts stem out of pure malignancy, camouflaged in a shroud of science. It is interesting that this sin is not the monopoly of the atheist camp alone. It has been practiced many times by the guardians of Theism as well. The scandal which has became known as the Shroud of Turin is one example. Herewith I will provide short summaries of some of these theories that have the only aim of denying the fact of the Big Bang. My aim is to enable you *not to be intimidated* by anyone who has marginal knowledge, but is snobbishly talking in scientific-like jargon.

THE STEADY STATE THEORY

This theory was first suggested in 1948 by three physicists who specialized in military radar. Herman Bundy and Thomas Gold came from Austria; the third was Fred Hoyle from Britain. They claimed that as the galaxies moved away from each other, new galaxies were continually forming in the gaps in between. They claimed that this keeps the whole Universe in a steady state at all points of space, and *at all times*, which means that there was no point of beginning.

To explain the source of the new matter that inserted itself between the expanding galaxies, they said that it is *continually created.* This theory required a modification of Einstein's Theory, modifications which could never be proved. In the 1960s, radio astronomy also showed that the distribution of the recent and old radio sources in the Universe strongly contradicts this theory. Moreover, the discovery of microwave radiation by Penzias and Wilson in 1965 also indicated that the Universe must have been much denser in the past. Thus the Steady State Theory simply couldn't stand the test of peer review, and was discounted.

THE THEORY THAT THE UNIVERSE IS YOUNGER THAN THE PROVEN AGE OF SOME OF ITS OWN STARS

A paper presented to a cosmological conference in 1995 described certain observations which claim to suggest a value for the 'Hubble Constant' higher than 50, (The Hubble Constant is a scientific value that is currently believed to be below 50). This would calculate the age of the universe no more than 8 to 10 billion years. So the universe would paradoxically be younger than the age of some of its own stars, whose ages have been unequivocally proved by other various techniques.

What looks interesting in this paper is the last sentence, which says *"unless a force of cosmic propulsion is speeding up the expansion rate."* This sentence is quoted from the book of Timothy Ferris titled *"The Whole Shebang"* page 243. The authors of the paper didn't wait to prove or disprove their claim. They were in a hurry just to throw doubt into the conference, before everybody goes home, and in order to catch the interest of popular astronomy and science fiction magazines. Understandably, this paradox soon made its way into headline news, and into laymen small talk.

THE THEORY OF AN EARLIER COLLAPSING UNIVERSE

This theory was presented by two Russian astronomers, Evgenii Lifshitz and Isaac Khalatnikov. In 1963 they suggested that the current expanding Universe resulted not from a Big Bang Singularity, but from an earlier contracting phase of a *previous* Universe. As that Universe was collapsing, the particles in it might not have all collided, but have flown past and then away from each other producing the present expansion of our Universe. After seven years of contradictory claims and experiments the two scientists simply withdrew this claim in 1970.

THE THEORY OF MULTIPLE UNIVERSES

This theory is the funniest of all the frantic attempts of the Atheist Camp. Their dilemma was double, first there was the Big Bang, and second there was the scientific data which were gradually building up a firm belief in the Universe being *tailor-made* for Man. The idea started as Science-Fiction. In 1937, Olaf Stapledon published a thriller titled *"Star Maker"*. It talked about different Universes with different physical laws, and even different dimensions of Space and Time. It was followed in 1950 by another story called *"The Wall of Darkness"*, by Arthur C. Clarke. In that story he imagined an endless River of Time which majestically flows through timeless space. It carries on its surface lots of foam-like bubbles, each of which is a separate Universe. Each bubble was started by a tiny point of 'mass-energy' which originated from the void, and exploded to form a full-fledged Universe. Each of these bubbles created its own constant values and physical laws. In certain Universes the properties of the matter and the physical laws *may happen* to be suitable for the existence of life. This will lead to the development of human observers who can wonder about their own creation.

In the 1980s, a biologist named C.F.A. Pantin fell in love with this fictional idea, and presented it as a scientific theory. He said that our Universe is not tailor-made for us, but it is as if we have entered a shop which has a large range of suits available in all cuts, shapes and sizes. We have just picked up *"off-the-peg"* the Universe which had the best fit for our needs as Carbon-based life. The multiple Universes exist within a wider infinite *meta-universe*, which is also in a state of inflationary expansion itself.

Pantin's whole theory was an outrageous offshoot of science fiction. In the 1980s the Theory of Multiple Universes replaced the Steady State Theory, which had fallen from favor in the 1960s. Curiously enough Pantin's theory was also mainly patronized by another *Soviet* physicist, Andrei Linde, from the P.N. Lebedev Institute in Moscow. However, in order to persuade lay people and some scientists the theory needed at least a faint flavor of scientific credibility, the patrons of the theory jumped into the boat of Quantum Mechanics. Here nobody will ever get certain knowledge as everything is in the realm of probabilities. This is exactly how malignancy can get shrouded in scientific camouflage. We know that the only contact between us and the vast reaches of our Universe is the electromagnetic waves. They include radio waves, light waves, microwaves, X-rays and Gamma rays, as well as the red-shift or blue-shift of them all. The only measure of distance we have in the range of millions and billions of light-years is the red-shift of the accelerating expanding galaxies in the Universe. So there is *no way* of communicating with any *other* Universe. This is the beauty of Pantin's theory, he will have planted the seeds of doubt in everybody's mind and heart, while no one will ever be able to prove or disprove what he says.

THE THEORY OF THE TAILOR-IN-REVERSE

With the disrepute of every theory the Atheist Camp adopts, another one soon pops up. None of them presents any new facts or discoveries. All share the same characteristics of speculation, theorization and imagination, wrapped in arrogant literary jargon and quasi-philosophical talk.

In 1995, a writer named John D. Barrow chose a brand new pathway. He argued that the likes and dislikes of the human taste have been molded by his Evolution, as well as by the laws of physics. In a book titled *"The Artful Universe"*, he is irritated by the expression of the Universe being 'tailor-made for Man'. He argued that on the contrary, *it is the physical laws which have tailored Man to suit his universe.* Even the publishers who were advertising the book didn't venture to say it is based on scientific facts, they say it is a literary work on the border of Science and Art.

As experimental and theoretical evidence mounted, it was finally proved in 1970 that the Universe had a beginning in Time, believed now to be at 15 billion years ago. Einstein's Theory and all the laws of physics were proved to be beautifully fitting to all the stages and phenomena of the Universe, starting from one minute after the Big Bang. But before that fateful minute, when the Universe was a Singularity, which means a tremendously dense single spot, the rules were beyond anything that we can comprehend, we should not try to ask questions about those mysterious sixty seconds, because there is no way to know.

CONCLUDING COMMENT FOR CHAPTER (4)

I mentioned on the first page of this chapter that after you enjoy reading it you will feel much closer to God. But there is a surprise comment in the last page, printed upside down.

The deeper we go into scientific facts, the more we realize the magnitude of intelligent harmony behind it all, a harmony which bespeaks the fact of an external force, an intelligent Creator who has initiated it all.

Unfortunately, fundamentalists, in all religions believe that the best way to keep Mankind closer to God is to let them know nothing but the scriptures. Not only that, but it must be their own particular interpretation of the scriptures.

In August 1999, a most bizarre decision was made in the USA, the country where most of the scientific research is taking place. The Kansas Board of Education voted 6 to 4 to remove Evolution and the Big Bang theory both, from the curriculum taught in the schools of the State. Why? They said that both contradict the literary interpretation of the Biblical Story of Genesis.

Which way would you feel nearer to God?
By myopic rigid interpretations,
Or by the glaring evidence of scientific facts??!!

CHAPTER 5

* * * * * * * * *

WHY *"GOD"* WON'T
SIMPLY GO?

THE TENACITY OF THE IDEA OF FAITH

* * * * * * * * *

"There are tens of different Creation Stories, one for every Religion. The theory of evolution merely adds another one. The main point of controversy is which Creation Story to believe.

In spite of all that, the principle of faith looks to be a *built-in code* in the human genome."

Dr. Hussein A. Amin.

"There is only one religion, though there are a hundred versions of it."

George Bernard Shaw.

First, let me define what I mean by the term *'FAITH'*. It is the un-explainable belief by Mankind that somehow every human is accountable for whatever he or she does or says during life. The questions of how will this accountability be executed, and in front of which court, differ from one ethnic group to another; and from one religion to another. But for every man or woman the belief is always there, deep in his or her psyche, even if he or she openly denies it. Even atheists may reluctantly accept their accountability, but their 'court' is social and cultural, on Earth, with no 'after-life' involved. Faith, as so defined, is comfortably compatible with every single religion, past or present.

Mankind invented the idea of writing spoken words only ten thousand years ago. Hence, the written history of Mankind extends no more than that period. At first, the written language was pictographic, that is symbols of scratches, geometric forms, or pictures of trees, birds and animals. Each symbol was a sentence with a full meaning. Later, the symbols became phonetic, each representing a sound and not a sentence. There are lip (labial) sounds such as M and P, and tongue (lingual) sounds such as L, R, S, T, etc.

Thirty thousand years ago, humans recorded their thoughts in the form of drawings on the walls of caves. People lived in caves because of the prevailing ice-age. Now we interpret their drawings as either religious prayers, or instructions for later generations. The sages recorded their observations of the sun, the moon and the stars, the solar and lunar calendars and their relationships to tides, female menstruation, and migration of birds and animals. They drew descriptions of how to catch prey, how to tame fire, etc.

Unlike animals, Man is capable of abstract thought. He has always pondered about the way he and the Earth were created, the mystery of life and death, and the dream of an immortal soul. ***International communications were impossible, and so for every isolated human group there was their own private speculation***. The various creation stories were tightly woven with ritualized religious beliefs. There are certain common characteristics shared by all those beliefs.

COMMON CHARACTERISTICS
SHARED BY ALL RELIGIONS

At the core of each religion there is a Creation story, which explains how the world began. There is always a *Chosen People*, with a story describing how these people arrived at the center of that belief-system. There is often a mystery and a set of sacred events and rituals. The rituals usually include prayers, (at specified daily or weekly times), some sort of fasting, (which means abstinence from bodily pleasures such as food and sex), as well as two or three 'feasts' every year. There are often certain times when some sort of meditation is practiced, commonly in a place perfumed by incense. Rhythmic movements or ritualized dances are common to some, but not all religions. In some religions there is a holy site where a yearly pilgrimage is observed, such as Jerusalem, Mecca, Varanasi (for Hinduism and Buddhism), and Mount Fuji (for Japanese Shintoism), for example. For the pilgrims it is a spiritual journey, for the locals it is tourism, bringing them financial benefits. Then, there is a system for punishment and reward in an after-life. The description of that after-life differs from one group of religions to another.

I believe it is relevant here to summarily review the spectrum of contemporary religious traditions. Their creation-stories are also summarized as described in the Encyclopedia Americana under the title 'Creation'.

SUMMARY OF TODAY'S MAIN RELIGIONS

There are four main religious groups in today's world: Monotheism, Hinduism, Oriental Religions, and Atheism. A fifth minor group consists of some examples of blind spiritual devotion to a person, rather than to a God.

The first *organized* religion in today's world is the monotheistic group of religions, which are Christianity, Islam, and Judaism, arranged according to the latest United Nations statistics of the numbers of their followers. There are more or less two billion Christians of different sects (or denominations), geographically spread all over the world. There are more or less one billion Moslems of different sects, geographically spread around the world. Judaism followers count only more or less twenty million, mainly because preaching Judaism to other people is not a common practice. The second group of organized religions is Hinduism, which is practiced mainly on the Indian continent, and claims more or less 700 million followers. The next largest group is the Oriental Religions, which claim roughly 1,400 million followers. They are subdivided into several religions, mainly Buddhism, Confucianism, and Shinto. The fourth group is simple straight forward Atheism.

Monotheistic Religions, share the following characteristics:

(a)   The belief that Man is formed of a mortal body, and an immortal soul.

(b)   During his earthly life Man has got to do good and avoid evil, as well as to worship his Creator, in preparation for an afterlife.

(c)   After he dies, his soul will face reward and punishment in an eternal afterlife.

(d)   The belief in the existence of a Creator, who created the universe, the earth, animals, plants and finally Man. Literal interpretation of the Scriptures claims that this Creation took six days to execute, and that God rested on the 7th day. It also claims that Creation took place only about six thousand years ago, and that at some time later a great Universal Flood covered the whole earth, drowning all the sinners.

Hinduism comprises several subdivisions which share the following characteristics:

(a)   The belief that Man is formed of a mortal body and an immortal soul.

(b)   The belief in the existence of a Creator for the whole universe including Man.

(c)   During his earthly life Man has got to do good and avoid evil.

(d)   When the mortal body is deceased its immortal soul gets re-incarnated into another human body, or the body of an animal. If the deeds of the deceased have been bad during his life, then his soul will be forced to enter into the body of a lowly creature or animal, or the body of a human of a lower Caste. If his deeds were good then his soul will enter the body of a human of a higher Caste. If the good deeds are repeated several times in successive lives, or if he lives and dies and gets cremated in the Holy city of Varanasi, then that soul is not born again into earthly life. It gets transferred into the eternal bliss called 'Nirvana'. This is simply, the Doctrine of Re-incarnation. The refusal to kill any animal in general and cows in particular, comes from the belief that these animals are animated by the souls of people serving punishment for bad deeds in previous lives.

The creation-story in the Hindu Scripture Rig-Veda tells about the marriage of Heaven and Earth, which begat Indra, the chief of Vedic Gods. He grew to an enormous size, separating Heaven from Earth.

Oriental Religions including Buddhism, Confucianism, and Shinto, share the following characteristics:
(a)   Man is formed of a mortal body, and an immortal soul.
(b)   On earth, Man has got to do *good* and avoid *evil*. In Buddhism, the Doctrine of re-incarnation is also valid.
(c)   The most important ritual in these religions is meditation. It requires absolute cleanliness of the body, and of the place, followed by silent meditation lasting as long as possible. Experienced meditators describe reaching a state of overwhelming ecstasy, very similar to the ecstasy experienced by the Sufi people in Islamic Sufism.

Oriental Religions differ radically from the two previous groups. It was decided by their founding sages that there is no way of trying to think of, or even contemplate the three main metaphysical questions: the Creator, the immortality of the soul, and the afterlife. The religion decided that these subjects are beyond the capacity of the human mind.   It is enough to teach people how to lead a good life on Earth. *"From Good Comes Good",* said The Buddha. Doing good and avoiding evil will offer happiness to Man's life and to the lives of others, but the repercussions of this on whatever supposed afterlife is beyond the capacity of human thinking, and should never be discussed or even contemplated. The Oriental Religions are not dogmatic; they allow their followers to adopt any other religious beliefs as well.

There are however, certain differences between the three main Oriental Religions. Buddhism, for example, favors the refusal of all forms of earthly pleasures and luxuries, including money and sex. On the contrary, Confucianism encourages its followers to enjoy the pleasures of life, provided they treat the others in the same way they would like to be treated. Confucianism also encourages the appreciation of beauty, music and arts. Confucianism also stresses loyalty to the family, and ancestors, rather than to the country in general.

In terms of creationist beliefs, the Chinese creation-story starts with the Cosmic Egg which broke apart. The lighter and brighter part formed Heaven, and the dark heavy part, the Earth. The Japanese Shinto creation-story tells of Heaven and Earth separating out of Chaos. The eighth pair of Brother & Sister Gods, (Izanagi & Izanami), begat the islands of

Japan. Izanagi purified himself in the sea water. The Sun-Goddess was born out of his right eye, the Moon-Goddess from his left eye.

Shintoism, practiced mainly in Japan, concentrates on the sanctification of their country, Japan, and the sanctification of the Emperor whom they call the Mikado. They believe that the present Emperor is the generation number 125 of the direct descendants of the God of Bright Sun who created the sacred land of Japan. The symbol of this God is represented in their flag. After the Second World War, General Douglas McArthur thought that the only way to demolish the warring spirit of Japan was to de-deify their Emperor, whom they considered God. The humiliation of their Emperor was a terrible blow to the Japanese belief-system. It has created a cruel spiritual vacuum for three generations so far.

ATHEISTS SHARE THE FOLLOWING BELIEFS:

(a)   Man is formed of a mortal body, and completely perishes on death.
(b)   There is nothing called the soul. The difference between dead and alive bodies is purely physical. Our present knowledge cannot define that difference, but one day science will clarify its nature.
(c)   There is no afterlife, and no Creator. Life emerged spontaneously by purely physical means.

Atheists are widespread throughout the world, with no particular geography or statistics. They comprise the following groups:

(a) Many scientists. The research and books of Charles Darwin became the first scientifically plausible arguments adding to this reasoning. The idea was given its famous name, the Theory of Evolution, and became the first *supposedly* scientific challenge to religious beliefs.

(b) Communists who *need* this idea, because they want to change the human nature of people into the nature of ants and bees. This cannot be achieved in the presence of any religious beliefs, and so those must be completely erased first, even through use of force.

(c) A Silent Majority of people who belong to all religions, and every human society and race. They have doubts, but do not speak them. The reason for this is twofold: First, is the apparent *scientific* appeal of evolutionist talks. Secondly, is the *Just-So* dogmatism of the teachers of theology, especially when it becomes extremist and fanatic, and intolerant to any argument. This approach drives people away from religion in general, rather than otherwise. A glaring example is the infamous Catholic Inquisition Courts some centuries ago.

RELIGIOUS FRAUD

In their over-enthusiastic zeal to control the minds of people, the clergy sometimes used twisted means. The scandal which has become known as 'The Shroud of Turin' is a good example. It is a length of linen that for centuries was purported to be the burial garment of Jesus Christ. It was first *'discovered'* in France in the 14th century, and later preserved in the Cathedral of Turin, Italy. The cloth showed the supposed faint image of Jesus, with markings that allegedly correspond to his injuries at the time of crucifixion. In 1988, the age of the cloth was finally determined by the method of carbon-14 dating in three laboratories, in three different countries. It was proved to have been made sometime between 1260 and 1390 A.D., conclusively proving that the Shroud could not have been Jesus' burial cloth. The idiot who fabricated that shroud may have thought it a good idea to increase the number of the believers. Instead, what he did was nourishing the seeds of doubt in the hearts of even the staunchest believers on Earth, when its fabrication was discovered.

Another religious forgery was the so called 'Donation of Constantine'. It was a supposed grant by Emperor Constantine to Pope Sylvester the First. It gave the Papacy strong authority over the Churches, as well as the State. It led to the well-known power-abuse by the Catholic Church for centuries. In 1440 A.D. it was proved to be a forgery. [Boorstin, 1998, page 68.]

*Evolutionist enthusiasts cite these and similar incidents as, **unfair**, evidence against religion in general, as well as the principle of faith per se.*

SCIENTIFIC FRAUD

Some amateur scientists in search of fame or money have willfully presented fabricated material and evidence. One famous example is the hoax of the 'Piltdown Man'. In 1912, Charles Dawson, an amateur geologist, announced his discovery of a cranial fragment and a jawbone in the area of Piltdown in South Sussex, England. He claimed they belonged to a prehistoric man half a million years old, carrying characteristic similar to those of apes. The case was labeled as the missing link between Ape and Man. The story thrived at a time when evolutionists and creationists were at an open war-of-words, without any real scientific data. Genetic science was still unknown. In 1953 the fossil was re-examined using fluorine dating methods. The skull was proved only a few hundred years old, and the jawbone proved to belong to an Orangutan.

In the year 2001, in a similar incident, a Japanese paleontologist named Shinichi Fujimura was caught red-handed while implanting artifacts at an archeological digging site, 350 kilometers north of Tokyo. For 20 years he kept on *'discovering and illuminating'* Japan's prehistory. In 1984, he was even welcomed into the ranks of Japan's Archeological Association. After being caught, he admitted to having fabricated everything. Japan's Universities were shocked by the hoax. The deception will cast a long shadow on Japanese science for many years to come.

*Clergy in all religions, cite these and similar incidents as, **unfair**, evidence against evolution in general.*

THE 20TH CENTURY STATE OF RELIGION

The 20th century started with two strong dogmas adamantly denying the principle of Faith. The first dogma was communism. Lenin preached that, *"Religion is the opium of the masses"*. In his view, religion was invented by the rich and aristocrats just to help the poor passively accept their own misery. In all communist countries, Christianity, Islam and Judaism were hunted and openly persecuted.

The second dogma was 'science', which reached its climax in the 20th century. It has found scientifically understandable explanations for all the phenomena of nature, previously unexplained. Thunder, lightning, earthquakes, droughts, plagues, all became simple earthly incidents with no supernatural forces involved. Soon, scientists started to question the very principle of faith. Many of them started to *symbolically* park their Faith, if they had any, outside the scientific laboratory doors. The phrase 'Is God dead?' became an arguable question. Coined by Nietzsche in the early 20th century, the phrase and its variations became commonly seen titles in laypeople magazines towards the end of the century.

*Suddenly, 'Religion' snapped back, and with a vengeance of its own. It rebounded towards the end of the century, and this time it had a dangerous flavor: a **'Fundamentalist'** one.*

RESURGENCE OF FUNDAMENTALISM, IN ALL RELIGIONS

The word 'fundamentalism' means the strict and literal interpretation of Religious Scripture, in a most orthodox way, with no lenience, tolerance or mercy. Everything became a "Just-So", no argument is tolerated.

In the mostly Christian west, and especially in the USA, it started with a school teacher called John Thomas Scopes. He was tried in 1925 for teaching the emerging science of evolution. In 1999, The Kansas Board of Education voted 6 to 4 to remove evolution, as well as the Big-Bang cosmology, from the curriculum taught in the schools of the State. The Board of Education's position was that both sciences contradicted with the literal interpretation of the Biblical Story of Genesis. It looks quite odd that in the country where most of scientific research takes place, statistics show that more than 55 % of Americans still believe that the Earth was created only six thousand years ago. A cut-throat debate is raging in the USA between evolutionists and creationists, for which some detail is given in next pages.

Besides mainstream Christianity in the USA, various fundamentalist cults crop up from time to time. In 1978, more than nine hundred people in Jonestown killed themselves by drinking poison at the order of their 'messianic' leader Jim Jones. He called his headquarters "The People's Temple", where he preached against the "Sky God" of Judeo-Christianity, allegedly used by the capitalists to justify the oppression of the blacks.

In a more recent incident, in Waco, Texas, USA, a cult named 'The Davidians' were totally brainwashed by their charismatic leader David Koresh. The female members of the cult were his sex-slaves. They all perished in a mass suicidal fire when Federal authorities tried to investigate their ranch. In 1995, a maniacal dissident, Timothy McVeigh, avenged their death by terrorizing Oklahoma City. He blew up a government building, killing about 200 people.

In October 1994, followers of a cult named "The Solar Temple" ritually burned themselves to death in two towns in Switzerland, and one week later in a suburb of Montreal, Canada.

In all these examples of *'home-made-religions'* there were certain common characteristics. The leader was always a charismatic authoritative demagogue, on the border of insanity. Each exerted unhealthy, unchallenged sway over his followers and the rituals he held were virtually brain-washing sessions. The willingness of the new recruits to abandon their good homes and societies was a troubling symptom of a profound   *spiritual-vacuum* that had to be filled. They were simply exploited by the first hypnotizing man that crossed their way.   This is a very important point about the *religious instinct* in Mankind, about which I shall deliberate more in Chapter 11.

Many Islamic countries were humiliated under both colonial and communist Empires. With the fall of both, religion made a fundamentalist rebound, similar to what took place in the West. Tolerance became a very difficult word to swallow, and the fundamentalist clergy found their chance to exploit the previously oppressed millions of people.

Fundamentalist Judaism took two forms. One was the romantic appeal of the creation of a special State for 'Jewish-ness'. This form is entertained by the Jewish communities thriving in Western countries, but still preferring to stay where they are. The other form of Fundamentalist Judaism took place in the created State itself. An ultra-orthodox minority could get a degree of political clout much larger than their percentage in the society, enough to steer the State in their own fundamentalist way.

In the Indian sub-continent, the colonial era ended in a miscalculated political blunder in a small mountainous area called Kashmir. Had it stayed as a purely political and territorial dispute, it could have been solved by now. However the issue got hijacked by religious fundamentalists on both sides of the dispute, and is now a potential atomic hell.

EVOLUTIONISM - VERSUS - CREATIONISM

Evolutionism versus Creationism has been a hot debate raging for several decades; particularly in the USA. It is no more a purely scientific discourse. It has become politicized and publicized in an unprecedented way. The weapons of this war are hundreds of magazines and books every year. Even the National Geographic Magazine got lured into the fray by its article titled *"Was Darwin Wrong?"* in its November 2004 issue. Both sides are vying to get public opinion approval, and consequently that of Congress members, and consequently billions of dollars in form of research funds.

Here are some examples of the fiery titles of some articles and books:
* *Fifteen answers to creationists' nonsense..!*
* *How to debate a creationist?!*
* *How to debate an evolutionist?!*

The aim in both "How to" books is not scientific truth, but mainly how to win a debate, thus becoming a higher paid lecturer!
* Evolution as dogma..!
* Creator, or blind watchmaker?!
* When faith and reason clash, evolution and the Bible..!
* *Darwin on trial.* A book written by Philip Johnson, who is a lawyer.
* Philip Johnson on trial..!
* Why creationism should not be taught in public schools?
* *Why evolutionism should not be taught in public schools?*

Additionally, the following statements are representative of this argument, [C] means creationists, while [E] means evolutionists:

[C] Evolution is only a theory.

[E] Relativity started as a theory.

[C] There are no intermediate forms, (between species), in fossil records.

[E] DNA comparisons can compensate for an incomplete fossil record, and one day intermediate forms will be found.

[C] Evolutionists are still fighting among themselves about many aspects of their own theory.

[E] Debate happens with every scientific subject.

[E] Creationists' alleged scientific claims can never get a peer-review validation.

[C] This is because we do not have enough funds, (!!).

[C] Mathematically, it is inconceivable that anything as complex as a single protein, let alone a living cell, could spring up by chance.

[E] *No comment...!*

[E] Bacterial resistance to antibiotics is a mutation.

[C] No, it is a manifestation of natural selection.

[C] Nobody has ever seen a new species evolve.

[E] By experimenting on fruit-flies we could produce reproductively isolated species in only 35 generations. They refuse to breed with other varieties. This is a sort of experimental proof of evolution.

[C] Mating repugnance and sexual isolation should not be the only definition of a new species.

[C] Argument that design necessitates a creator.

[E] Evolution over millions of years *can* produce design.

[C] Argument from personal incredulity.
[E] Personal incredulity has no value in science.

After that presentation of this array of conflicting statements, I find the following quotation to be most interesting. It illustrates how scientific facts are willfully manipulated by both sides of the infamous debate: [Dawkins, 1999, page 207]. He says:

> "Stephen J. Gould's ideas, about the Cambrian explosion of multi-cellular life are heretically against the basic principles of the Theory of Evolution. ***But he should not be publicly criticized, because at least he is on our side against the creationists.***"

Finally, the war has gone to the Federal Courts. Should teachers in schools teach their children about the Bible or about Evolution? There, the lawyers on both sides do not talk much science. The tactics they use are the same they use in criminal courts, which reminds me of an old anecdote:

> "One day a famous lawyer was hastily chauffeured to the court, hardly having enough time to read the case papers except for a few minutes during the trip. He presented a very strong case against Mr. A and then took seat. His client whispered to him in indignation, how dare you defend my opponent, Mr. B? Without any comment the lawyer picked up some blank papers pretending to read them, and stood again: Your honor, what I've just said are the flimsy arguments of Mr. B against my client Mr. A. Now I'll tell why this argument is false. And he won the case."

All the defenders of evolution or creation want from Federal Courts and State Boards of Education is to be able to get 5 to 4 verdicts that vindicates one side one year, and then the other side a next year. A recent more thorny issue is why talk only about the Bible, while there are several other Religions observed by millions of citizens. What is really at stake is not religious zeal, it is billions of dollars of, governmental and private, research-funds that are potentially the great prize of the winning side.

Does this debate interest me and you? Yes, and of course a big yes. This is because the outcome of this debate centers on Mankind, his origin, and his relationship to other forms of life. But let us talk about it in a different way, *a scientific way*. Let me crystallize the two warring viewpoints in a few pages.

SCIENTIFIC RULES OF THE DEBATE

The main source of 'wisdom' 500 years ago was Religious Scriptures. Their interpretation was monopolized by the clergy, of *all* religions. This gave the clergy both power and wealth, which are two *human* ambitions very difficult to relinquish. They literally 'appointed' and crowned the kings and rulers.

Egyptian Pharaohs were virtual Gods from the farmers' point of view. But still when one of them, Akhenaton, challenged the authority of the clergy they poisoned and replaced him. For several centuries the Catholic Church in Rome enjoyed legendary wealth, as well as legendary hegemony and authority over most parts of Europe. The feud between King Henry and the Roman Church was not about right or wrong. Even the King's desire to divorce his wife, Catherine of Aragon, was not the real issue. The basic issue was the authority which the Roman Church enjoyed over the whole of Europe. Henry's revolt was setting a bad precedent, and had to be fought.

The prevailing wisdom was that Mankind and his home, the Earth, are the center of the whole creation, including the stars. The Earth was a vast flat garden. Everything on it, scenery, animals, birds, flowers etc, had only the purpose of pleasing the eyes and body of Mankind, the children of God. The clergy were the guardians sent by God to look after His children.

Why did the Children of God tamely submit to that ruthless custody, which simply exploited them? Because Mankind *is* genetically religious, but most of them are ignorant enough not to know how to fulfill that *genetic* urge. *They think that by submitting their freedom and their money to the clergy they have done what they should do.*

The only other source of wisdom at that time was the vastly imaginative mind of Leonardo Da Vinci, who lived from 1452 to 1519. His scientific observations were far in advance of his times. His sketches predicted even future modern city planning, motor vehicles, flying machines, submarines, the armored war tank, etc.

Then the great shock came. Copernicus, by his calculations, and Galileo, by his telescope, suggested the Earth is neither flat nor the center of the Universe. Both paid dearly for their courage. The next big shock was what microscopes revealed: germs *are* the cause of illness, and *not* the powers of Evil. Sick people used to ask for prayers of the clergy against the evils that cause sickness, because they knew nothing better.

From that time on, scientific wisdom started to successfully wedge itself between Mankind and their self-appointed custodians. The wedge grew by the day, so much so that in the 20th century it is becoming unfashionable to talk religion, this is the other, mistaken, extreme. Unfortunately it is depriving the human heart of its genetic religious urge, and it is mainly the fault of the clergy, in *all* religions.

How do scientists perceive their wisdom? Scientists say that there are two main questions at stake. The first question is the *why*? Why the creation, life, death, happiness, misery, etc.? Scientists indignantly shun that question, and delegate it to the artists of talk, the philosophers. Philosophers can ponder these questions from now till eternity. The second question is the *how?* How we came to be. Scientists view that question as fair game. However, there are certain rules for that game of science. The aim is to avoid fraud, and prevent quacks from masquerading in white coats and sneaking into scientific shrines. The rules are mainly as follows:

Rule (1) says that we have only five senses. These senses are the ultimate Court of Appeal for any scientific evidence. Some may argue that we do not see sub-atomic particles, and yet we accept them as fact. This is just playing with words. We have come to prove most of the laws of physics and chemistry which turn the wheels of the Universe. We must believe when these laws tell us that there are black holes possessing tremendous powers of gravity or that there are subatomic particles held together by the four natural forces.

Rule (2) appoints a second Court of Appeal, and that is peer-review. All scientific evidence or experiment must be able to be verified and confirmed by peers: in journals, conferences and in books and other publications.

Rule (3) excludes any supernatural explanations for any theory or evidence. Scientists call them 'Science-stoppers'. Suppose we come to a riddle which we cannot solve. If we then simply say it is done by Divine power, then this is end of story, nothing more to say. There is a historical reminder of such a situation. The great scientist, Sir Isaac Newton noticed that some of his calculations about planetary orbits were embarrassingly flawed. They didn't fit well with the known data of stellar orbits. He then made the worst blunder of his life: he resigned the difference to Divine intervention. *"God would intervene from to time in order to correct the stellar orbits."* he argued. Later, when Laplace provided better calculations, God's intervention was no more needed. The incident was one big mistake in Newton's scientific career. If Mankind claims 'Divine intervention' at any point in scientific research, then there will be no progress at all.

However, this does *not* mean that science has no place for God, as some scientists say. God has, *somehow,* privileged Mankind by an inquisitive mind. We have to use that mind, that gift from God. We have to keep using it to the extreme, against any apparent puzzle, riddle, or seemingly unsolvable questions of *how?* It is by this third rule that science has progressed to its present peak, and will always do so.

Now, let us go back to the case of creationists versus evolutionists. By definition, creationists can present no scientific evidence, simply because everything is ascribed to Divine intervention. The only 'science' available to them under the big term of *'scientific-creationism'* is to criticize evolutionism, and make a public-relations issue of the flaws and weak points in evolutionist theorization. *'After all it is a mere theory'*, creationists will ultimately say.

Creationists' only documents are the Religious Scriptures. But *alas*! There are at least five major Scriptures, with sometimes contradictory information. Each is sanctified and observed by millions or billions of Mankind, so where or how to start? By the mere preferential talk in favor of any of those Scriptures, creationists are sure to meet totally blunt ears in at least two thirds of humanity.

By definition, scientists in general, and evolutionists in particular, will deny any Divine intervention in the mechanism of the Universe once it was started at the moment of the Big Bang, 15 billion years ago. They may be able to swallow Divine creation, on one condition, that God set all the rules of both Universe and life at the fraction of a second of the Big Bang, and afterwards He simply let go. This principle is called *Deism*. At the fateful moment of the Big Bang certain physical laws were validated. In Chapter 4 we saw how beautifully the physical world has materialized to form the Universe, down to the creation of planet Earth. Throughout time, it was guided and ruled only by those physical laws. This perfectly fits with the principle of Deism.

However, can the principle of Deism also deal with the creation and evolution of life and Mankind on planet Earth? Yes, it can with the stipulation that the process should also obey the same rules. Namely, being only guided by the same physical laws validated 15 billion years ago. This will be discussed in detail in Chapter (7).

IS GOD STILL "LISTENING" TO PRAYERS?

The principle of Deism scares every clergy on earth: *his prayers are simply doing nothing and there is no use for his services.* Clergy do not say so, instead they appeal to the public-relations tactics. They tell people that if they accept the principle of Deism, then their prayers are useless, and everything is determined in advance and doomed. On the other hand, staunch evolutionists argue that people who may merely contemplate the idea of 'God' are just superstitious minded.

This argument and counter argument is merely fruitless playing with words. Words are the playground of philosophers. I believe that Mankind can, and *must*, investigate the intricacies of the wheels set in motion by God, but his mind can never perceive the *"MIND-of-GOD"*, if we may use that term, first coined by the famous physicist Stephen Hawking. Our five senses are too short for that. However, an omnipotent God who has created all this from a pinpoint singularity some billion years ago is capable of dealing with His creatures *as He wills to act.* Neither scientists, nor evolutionists can chain His *"hands"* by whatever theorization they do.

He has privileged us with a trouble-making, inquisitive mind, as well as a consciousness that is different from any other creation. Both mind and consciousness have led us in less than 100 earthly years into what we are doing now, diving to the deepest seas, and flying to the moon. Once we contemplate all this, there is only one conclusion. Our mind *and* consciousness must have a reason and a purpose, both of which we are simply not equipped to know.

Still there is more to say about the listening God. It has been documented beyond doubt, in scientific and medical journals, that belief in *'faith and prayers'* **does help many patients to improve.** It adds greatly to their chances whether the disease is infection, injury or cancer. Their immunity system is stronger and they are less prone to defeatism, fatal depression or despair. We know that the immunity system is strongly affected by whatever goes on in the mind. If this is so, and if it is universal in all human societies, then it must be genetically-based in a way we do not know yet. Thus, we can safely say, on scientific grounds, that God is still listening to the prayers of people who do. God is a good friend to have; and curiously enough we are genetically prone to love that friendship.

I defined the term **Faith** *at the beginning of this chapter. In spite of all the upheavals in human history, the principle of Faith and Religion has shown a remarkable tenacity and resilience. This is simply because it looks to be a built-in 'code' in the human genome. Recent research in that direction is discussed in Chapter 11.> > > > >*

The principle of Faith and Religion is a built-in code in the human genome. Hence, feelings of respect for a supernatural Deity is almost universal for Mankind. Those noble human feelings were unfortunately exploited by the clergy in all organized religions. **They simply set themselves out as "God's agents"!** *I believe that this was, and is, the original root of the falsely called 'religious' wars.*

Figure (8)

A LOVELY COUPLE OF MIGRATING BIRDS
Their mother didn't tell them where to go,so it must be 'encoded' in their genes,
isn't it?

CHAPTER 6

* * * * * * * * * *

Co-Evolution,
And Adaptation

* * * * * * * * *

When philosophers or scientists try to convince other people of an idea, of which they themselves are not sure, they commonly use a certain gimmick: they label it by **giving-it-a-name**. Following that, they talk in favor or disfavor about the *name*, without going into detail. Two of the most common give-it-a-names are *Adaptation*, and *Co-evolution*.

In evolutionists' reasoning, the random, or haphazard chance, is the weakest argument of all. Evolutionists have provided enough scientific evidence for the process of Natural Selection. But Natural Selection alone can only improve the qualities of an *existing species*. The action of natural selection in this regard is called *Micro-Evolution*. This is different from creating a new totally different species, for example changing an ameba into a frog, a cat into an elephant, or a bat into a bird. Creating such new species is called *Macro-Evolution*. For macro-evolution to take place there must be genetic changes or mutations, on which Natural Selection starts to act, and ultimately get these *Macro-Evolutionary* products in several thousand or million generations. This will be how a totally different species gets created. In order to avoid any theoretical possibility of God's interference, evolutionists stipulate that these mutations take place by *pure random chance, in a blind purposeless manner.* In evolutionists' reasoning, there are lots of loopholes and weak arguments. They defend them by arguing that even Newton's and Einstein's theories had similar loopholes at their start.

In the following pages I have selected only a few examples of real-time life. Scientifically speaking, none of them can get fulfilled under the principle of random chance within the time allowed, which is the age of life on Planet Earth. They are: Chlorophyll and Hemoglobin, The start of sex, Echolocation, Parasitism, Symbiosis and symbiotic pollination, Camouflage and mimicry.

THE MOLECULES OF LIFE, CHLOROPHYLL AND HEMOGLOBIN

Chlorophyll is the cornerstone of all plant life. It uses energy from the Sun and transforms it into plant tissue and atmospheric oxygen. Hemoglobin is the cornerstone of all animal life. It is the only means by which animal body cells can breathe. A hemoglobin molecule consists of four chains of amino acids twisted together. Each chain consists of 146 amino acids, arranged in a particular specific order. If this order is changed the molecule *ceases to be a hemoglobin molecule,* and won't be able to carry out the function of oxygen transfer. *"The Encyclopedia Americana"* [1994, vol. 3, page 758] is quite clear about this point:

> "One protein differs from another protein only in the order in which the amino acids are strung together. Only twenty amino acids make up most of the protein in animal bodies. When these amino acids are connected in proper sequence, (like the words in a sentence), the protein will be a complete functional protein. If, however, one amino acid is deleted or changed, the protein becomes either a different protein, or a nonsense protein. One such mistake in the hemoglobin molecule results in the disease called sickle-cell anemia. It is a disease caused by the substitution of one amino acid called valine for the normally present amino acid called glutamic acid at a single location in the hemoglobin molecule."

We know that there are 20 different amino acids commonly found in living things. The number of possible ways of arranging 20 kinds of anything into chains 146 - links long is an astronomically large number, 1 with 190 zeros after it. A billion is a 1 with only 9 zeros after it. Such is the chance probability of happening to hit upon the hemoglobin molecule *by luck.* Computer calculations have estimated the time needed for such a probability to be billions of years, even longer than the age of the Universe. If billions of years are needed to create the hemoglobin molecule by random chance, then how many would be needed for creating the awesome complexity of any living body. The chlorophyll molecule is even more complex than hemoglobin. And 3,500 million years ago the blue green algae had to acquire it *by pure chance,* in order to be able to fill the atmosphere with oxygen.

In his book, *"The Blind Watchmaker",* pages 45 to 49, biologist R. Dawkins admits the astronomical improbability. He counterclaims that the hemoglobin molecule has been built, not as a single shot, but rather step by step by gradual *'cumulative-selection'.* This is one of the frequent Just-So stories of evolutionist literature, not much different from the Just-So stories used by clergy. The expression *'Just-So'* refers to any dogmatic statement for which no argument is tolerated. The word 'cumulative-selection' is a very transparent claim that can easily be seen through. The hemoglobin molecule is a protein with a specific order of amino acids. With any single change it ceases to be a hemoglobin molecule, *it is 'All-or-None'.* If

Mr. Dawkins claims gradual cumulative selection in this case, then he is unknowingly acceding to a purposeful, directed, step by step selection towards a *PRE- determined* goal, rather than a haphazard chance occurrence. This is supposed to be heretical, in the evolutionists' world.

THE START OF SEX

Single-celled organisms monotonously ruled the Earth for three billion years. They multiplied by simple division into exactly identical organisms, generation after generation. Five hundred million years ago, the Cambrian Explosion of multi-cellular forms of life took place. Suddenly, there appeared another way of multiplication which is sexual. The organisms got differentiated into males and females, each contributing a sex cell with only half the number of chromosomes to share in the genetic structure of the offspring. This allowed variety and differences to gradually accumulate in generations of offspring. Natural selection could then exert its influence to let the organisms adapt to environmental changes. Which is male and which is female is defined by a simple character, namely the number of sex cells produced by each. Males produce thousands or millions of sex cells, pollen in the plant kingdom and sperm in the animal kingdom, in an almost un-interrupted stream. Females produce only one or two sex cells in each sexual cycle. When the male and female sex cells unite, a new individual is started afresh, with two complimentary half-sets of genes.

It takes two individuals to agree to such a plan before starting to implement it. Further, it takes two individuals to have intercourse and exchange sex cells. The necessary instructions had to be inserted *in advance* into the genes of the individual who will become male, in order to develop a cell with only half the number of its genes, as well as a specific *masculine* gene. This procedure had to be *simultaneously* complimented by a similar set of genetic instructions into the genes of another individual who will become female. She will develop a cell with only half the number of her genes, as well as a specific *feminine* gene. Such an arrangement smacks of being made by an outside force, other than the males and females involved.

It is really interesting to see how evolutionist books hastily cross over this embarrassing area. In the *"Blind Watchmaker"*, this very crucial step in the theory of evolution was humiliated into a few words on page 268 with no explanation, [Dawkins, 1991.] In *"The Red Queen"*, a book fully dedicated to sex and its relationship to evolution, [Ridley, 1993], the only few words explaining that important item were on page 28 and read as such:

> "Why must sex have a purpose, may be it is just an evolutionary accident that reproduction happens that way,   like driving on one side of the road ". (?!)

THE HIGHER SENSES, ECHO-LOCATION

Touch, taste, smell, vision and hearing are but examples of the degrees of sophistication a living body can reach. In these five senses the individual passively acts as the recipient party. Light waves get reflected off the surface of any object and are passively received by the eye. Sound waves get carried in air or water and bombard the ear drum. Odors travel in molecular traces in air or water and reach the nose. Tasty elements directly affect the mouth and tongue. Lastly the skin feels whatever actually touches it, but nothing otherwise.

Clergy of the 18th century were stuck with the eye-design (camera-like) idea, which bespeaks of a designer Creator. Darwinists found easy answers in the light-sensitive spots on the skin of some fishes. These, they argued, could easily evolve by natural selection into eyes, in the course of a couple of hundred million years. I won't be dragged into that dreary old debate. I'll talk instead about a very clever sixth sense, which is present in only a few life forms, and that is *echolocation*. We find this sophisticated sense in bats, dolphins, and whales. Bats earn their dinners in the dark of the night. Dolphins and whales find their way in miles and miles of the under-water world, where light can be of little help. In this legendary means of communication, the animal is not on the passive recipient end. On the contrary, it actively initiates certain sonic or ultrasonic waves, and receives them back carrying all the information it needs. For the reader who is not familiar with the technicalities of radar let me give a brief account.

Radar devices emit radio waves, which get reflected off any object in their way. The receiving part of the device analyses the echoes of the returning waves, first to judge the distance of the object, and also to judge its shape. Sonar devices act in similar way, but emit sonic or ultrasonic waves instead of radio waves. Ultrasound waves travel at the speed of sound, 300 meters per second, in air, but radio waves travel a million times faster, at the speed of light, 300,000 kilometers per second, in vacuum.

In the Second World War, both warring sides heavily relied on both sonar and radar in submarine and air combat. In peace time they help detect schools of underwater fish, sunken ships etc. In medicine they are an important diagnostic means, such as examining an unborn baby in its mother's womb.

Sound waves are low pitched and they have long wave lengths. Their echoes will not resolve the differences between closely spaced objects. Ultrasound has much shorter wave lengths, and its echoes can give a higher resolution and more defined pictures. The waves get emitted in pulses; this allows the receiving end of the device to separately interpret each pulse.

The pulse frequency can be adjusted to the needs of the operator. If nothing particular is visible then the frequency can be kept low, to save energy and wear & tear of the machine. When something gets detected, especially if it is moving, the frequency can be increased so as to continuously update the echo-pictures received, even up to 200 pulses per second.

The emitted ultrasound waves need to be really loud to start with. Sound waves quickly fade away, in proportion to the square of the distance from the source. The reflecting object acts as a secondary source for the returning wave, which further quickly fades down. The receiving end of the device needs to be very highly sensitive in order to detect and analyze the faint echo waves it ultimately receives. This can pose a serious hazard if it gets exposed to the extremely loud originally emitted waves. Thus, in modern sonar devices a technique called *send / receive switching* is applied. The receiving part is switched off at the exact fraction of a second of pulse emission, and switched back again in time to receive the echo. This is very much similar to your heart muscle, which once stimulated to perform a beat becomes refractory to any further stimulation until it has finished the ongoing beat.

A further perfection of radar and sonar devices is called *chirp-radar technique*. The emitted wave starts at a high pitch, which then changes to a lower frequency before it stops. This technique helps the receiving end of the device in its analysis of the echo. It helps also in differentiating the returning echo from other echoes caused by other sonars in the vicinity.

The enemy in a war may resort to *jamming techniques* in order to neutralize your device. They may emit false echoes, or try to emulate and nullify the frequency and pitch of your waves. This can be counteracted by *coding* the frequency and pitch of your device so that it becomes comparable to a finger print that can be identified even among hundreds of other echoes. Such a private code could also be frequently changed from time to time for further security.

Sonar and radar waves also help their interpreters in estimating the velocity of the objects they reflect upon, and that is by the well known *Doppler-shift*. The sound of an approaching ambulance siren is of a little higher pitch than the real sound of the siren. When the ambulance is moving away the opposite takes place. Doppler shift techniques can be utilized in all forms of waves, sound, ultrasound, radar, or light waves. It is also employed in the sophisticated guided missiles. It enables the missile to keep on the trail of a fast moving object trying to escape its chase.

THE FLYING RATS

To develop these highly sophisticated echolocation systems, it cost Mankind two devastating world wars, as well as billions of dollars in scientific experimentation and research. In 1940 two American zoologists, Donald Griffin and Robert Galambes, astonished the National Zoology Conference by their first report about "Bat Echo Location System". They had spent hundreds of hours investigating and experimenting on how bats find their way in the dark. Some of the participants in the conference became indignant. They couldn't believe that such highly classified military secrets could have been casually utilized by these flying rats for millions and millions of years.

The two scientists proved that bats have mastered all aspects of the echolocation business. They emit the ultrasonic waves through the nose or mouth, and receive their echoes by a highly specialized ear apparatus. They have grotesquely shaped appendages hanging from the nostrils and ears. Griffin showed that bats use certain minute muscles to switch off the ear at the fraction of a second the impulse is given. This is the same as the *send-receive* military technology. He also proved that bats utilize all the tricks of *chirp-radar, frequency-modulation, frequency-coding,* etc. They were also able to use the Doppler Effect in estimating their own speed, or the speed of their prey. They easily find their way in pitch dark caves even in the presence of thousands of other bats, each working its own sonar machine. A female bat can home in, and feed her little baby, amongst thousands of bats in a cave with the biggest ever confusing and deafening ultrasonic noise. I am sure that this is a feat which could defy the best human military engineers. The echolocation systems of dolphins and whales have also been extensively studied, but it is enough to concentrate on one example.

Echolocation is an *active* higher sense, unlike vision and hearing in which the animal is a passive recipient. To start such a system from zero it requires a lot of experimentation and active study, not only of the environment, but also study of the habits of predators and prey that may get in the way. Natural Selection can improve the system if it is *already* present. For example, if butterflies successfully evade a certain bat with deficient technique, this bat will die young before producing offspring. Only bats with faster and more accurate equipment will get their dinners and will survive.

But the Genes for that equipment have got to be there to start with. These genes will manufacture the necessary features of an echo-locating animal, for example changes in the mouth and nose, changes in the outer ear flaps, changes in the receptive ear, and above all, changes in the brain which will compute all the signals received. As such, these are a very clever and coordinated set of genes. According to the evolutionistic literature this set of genes must have been initiated together as a lump sum group by the very lucky strike, of a purely random chance-mutation.
A really lucky throw of dice!

GENETICALLY-IMPRINTED INTER-SPECIES COORDINATION

This area is a very weak argument in evolutionist literature. I have chosen only a few examples to illustrate my view.

First, is an elaboration on about how scientists got all the information which we are going to present. We used to describe an animal, insect or plant, by its external appearance, by its feeding habits, and by the number of eggs, fruits, or offspring. This is not acceptable anymore. Methods of biological research have become much more advanced.

Today's photographic equipment is highly sophisticated indeed. We can fit a camera in front of a flower or a beehive for round-the-clock photography, for days and weeks. Fiber-scope cameras can be made in miniature or even microscopic sizes. They can be introduced into the burrows of rodents or worms. They can even be fixed inside the gut of an insect or animal, or in between the petals of a rose. Microscopic microphones can betray the private lives of many insects and animals. We can study an insect one millimeter long on a wide screen, and watch it in detail, recorded in full color and stereophonic amplified sound. Research teams equipped with such sophisticated gear have reached every corner of the world. Scientists have spent months and years in forests, marshes, and deserts, studying all forms of plant and animal life. In the last ten years, a great wealth of information has been gathered. Bits and pieces beautifully fell together like a gigantic jigsaw puzzle. All this has revolutionized our knowledge of the way every living being reproduces, lives, and feeds.

I became fascinated by this new science, and have been able to obtain and read the latest of texts, pictures, and videos, from the four corners of the world. My particular interest has always been the manifestations of genetically-imprinted interspecies coordination.

THE FUNNY STORY OF HYDATID DISEASE

In the year 1960 it all started by a funny incident of hydatid disease. In this disease, single or multiple cysts develop in various organs of the patient's body, and its mortality is very high. The cyst acts like a tumor, but is caused by the parasite called echinococcus granulosus. I learned its pathology as a medical student half a century ago. It leads a two-cycled life. In one cycle it is a *Dr. Jekyll* -like innocent hermaphrodite worm. It lives peacefully as a harmless parasite inside the guts of dogs and wolves, which act as its primary host. Its eggs get passed out with the excrement of dogs and wolves. The eggs settle on the grass, and ultimately reach the stomachs of grass eaters such as cattle and sheep, which act as their intermediate host. Humans get similarly infected by the excrement of dogs through intimate contact with dogs, as in shepherds, and also in children playing with their pet dogs. In this case, Man also becomes an intermediate host.

Once the egg reaches the stomach of an intermediate host it starts a very aggressive behavior. It sheds its coat, and becomes a wicked *Mr. Hyde*. It stabs the wall of the stomach to reach the veins. The blood stream carries it to important organs such as the liver, kidney, heart, brain, or even bones. Each egg holds tight to these new sites, and starts to grow into a different form. This new form is a hydatid cyst, lined by thousands of germ cells called scolices. Each scolex is able to create a new *Dr. Jekyll*-type worm, if only it could reach the stomach of a wolf or a dog. These germ cells achieve their aim by wicked means; they *kill* their intermediate host, and get it eaten by a wolf or a dog. The hydatid cyst destroys the sensitive organs in which it grows and the health of the animal deteriorates until it ultimately dies. A quicker way to kill its host is for the cyst to burst open; this will release its highly allergenic fluid content; an anaphylactic shock takes place and immediately kills the intermediate host. A third means is for a hydatid-infested bone to break, and then the animal becomes an easy prey. Whether helpless, or dead, it is ultimately visited by Mr. Wolf, who enjoys a free meal, for which he made no effort at all. During his meal he takes the cyst and its contents as well. The scolices reach his intestine; they hatch, and rapidly grow into adult echinococcus worms. *Fait Accompli!*

There is no other way for this circuitous life cycle to get fulfilled. If the eggs of the worm get put into the mouth of their original host, wolf or dog, or any other carnivorous animal or fish or bird, it simply perishes. It neither develops into a Dr. Jekyll nor a Mr. Hyde. If the germinal cells of the Hydatid cyst reach the stomach of any animal other than wolves and dogs, it will perish and will not develop into an adult peaceful echinococcus worm. The whole circle is a specialized drama for three actors. The three actors are canine carnivores (wolves and dogs), echinococcus worms, and herbivorous cattle and sheep, (and coincidentally also Man). During my undergraduate and postgraduate studies I wondered about a possible meaning for this three-sided drama, but I didn't follow that lead any further.

I was surprised to notice the *"Encyclopedia Britannica"* wrongly describing the life cycle of this worm. It says that *"it is a tape worm common in sheep, cattle, camels, dogs, and many other mammals. The disease can develop in humans upon ingestion of the eggs"* The story is scrambled without differentiating between the primary and the intermediate hosts. Surely no physician was consulted about this statement. [E. Brit. 1995, vol. 4, under title Echinococcus.] However the *"Encyclopedia Americana"* correctly describes this life cycle. The entry is signed by an M.D. physician, which explains the correct description. [E. Amer., 1994, vol. 14, under title Hydatid disease.]

I spent half my surgical career in the city of Kuwait, as a Consultant Surgeon, for 22 very interesting years. Out of a total of eleven thousand operations there, I operated on more than a hundred Bedouin shepherds having Hydatid cysts in various parts of their bodies. I told them how to avoid that infestation again. I twice published popular articles in laymen journals warning parents about the danger of having their children infested through pet dogs, if the dogs are allowed to roam outside the house eating dead carcasses of sheep.

One of my patients was a sixteen years old student with a ruptured Hydatid in his spleen. His grateful father was very impressed by the appearance of the removed cyst. At a cozy moon-lit garden dinner he told many memories of his eighty years. He told me that he has witnessed an exactly similar cyst in the broken leg of one of his sheep a few years earlier. He was attracted by the voices of its helpless battle with a wolf. He scared the wolf away and slaughtered the wounded sheep. *" ..This wolf, doctor, has indirectly used the wicked eggs of his intestinal worms to catch my sheep, and eat it ..."* he further said. He was precisely correct.

CARNIVORES VERSUS HERBIVORES

On that moon-lit evening of 1960, it suddenly flashed in front of my eyes. This sort of parasitism is one of thousands of means which keep the balance between carnivores and herbivores and various creatures on this Earth. I looked up the title of "Parasitism" in every book or Encyclopedia available in the sixties. The only explanation offered was that it is an *adaptation* of one living being to the life inside the body of another living being. What adaptation? How did it get inside the other body to start with? The life cycle which we have just explained is a fully organized drama that has been going on for millions of years. It requires definitive characteristics and behaviors, with clear cut *genetic* instructions, which have got to be *simultaneously* complimentary in all three groups of actors. If ever we could accept that one of them has *evolved* that way, how could we accept that the other two have *so conveniently* evolved in that astonishingly co-operative way?

Evolutionist books simply explained it away by their famous tactic of *give-it-a-name*. They coined the name *Co-Evolution* as their only explanation. But this is just playing with words with no scientific explanation. The only *comical* translation of this word of co-evolution in this context is as follows: At one time, in the caves of history, millions of years ago, a summit meeting was held between the ancestors of these three species, during which they wrote and signed this pact.

Then every one of them dutifully, and indelibly, inscribed the terms of the pact into his descendants' genes.

I gave credit to the little worm called echinococcus granulosus by telling its story first. I discovered that this little worm is a mere single example of thousands of similar fascinating dramas. Each example is an intertwined life cycle that binds two or three or more species in an eternally balanced dance. Some of these miraculously beautiful examples are herewith presented below. You cannot resist getting deeply touched by the graceful harmony of them all. Evolutionists look down at any *argument from design*. But this is not the design of an organism; it is a tripartite genetic arrangement of three organisms, beyond any random happenstance.

OTHER EXAMPLES OF PARASITISM

It is interesting to know that a full ten percent of all known species live as parasites on the bodies of other creatures. However we will describe only a few, the hydatid, just described, as well as malaria, Bilharziasis, and the tape worm of the European song-bird.

MALARIA

Here there is also a tripartite arrangement between three different species, the malaria protozoon, the Anopheles mosquito, and Man. Each of them has complete knowledge of the anatomy, physiology, and social habits of the other two. Man got this knowledge by using his mind, and creating scientific knowledge. But the other two species got this knowledge imprinted into their genes. They act on that knowledge in their successful campaign, taking advantage of their common host, Mankind.

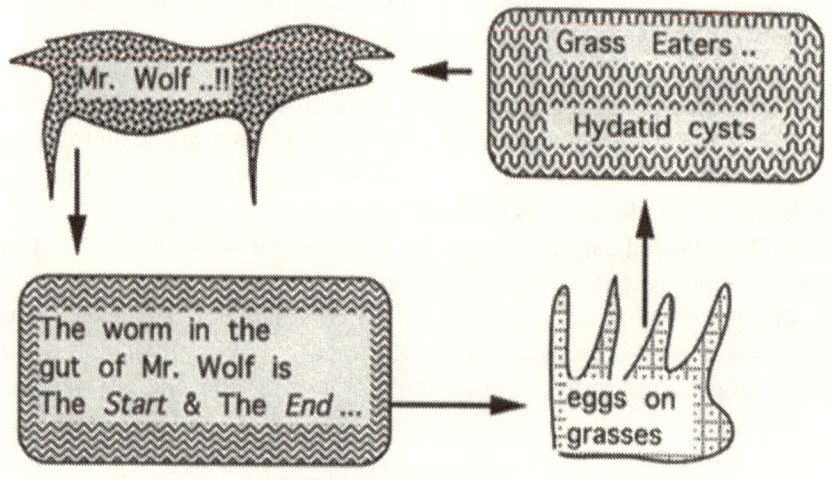

Figure (9) The cycle of hydatid disease

The malaria protozoa attack the red blood corpuscles in the human blood, feeding and multiplying inside them. The red blood corpuscles ultimately rupture, releasing thousands of young parasites into the blood stream, as well as their waste-product toxins. The protozoa make a point of having these ruptures taking place all at the same time, in a collective way. This maximizes the effect of the toxin on the patient. He gets a sudden shaking chill, followed in a few minutes by a very high fever and sweating. This always takes place at night, and night is the working hours of the mosquito. It gets attracted to the humid hot body of the patient, who is usually stuporous enough not even to feel its sting. The fever causes the skin surface to flush with dilated capillaries; the mosquito senses the warmest part and gently lands there.

Once the mosquito's needle reaches a blood vessel, it injects a heparin like substance to prevent the blood from clotting. While getting its fill, many of the freshly ruptured young protozoa take the chance to get on board. They fasten their seat belts, and get ready for the flight into another patient. When the mosquito reaches another patient, and while injecting the heparin like substance, many of the protozoa unlock their seat belts and come down before the mosquito takes off again, thus landing into a new miserable host.

It is worth noting here that the needle of the mosquito looks very much like the needle commonly used by several heroine addicts in a single session. It would theoretically transmit Aids, typhoid, and all sorts of viruses and bacteria. But this never happens as all of these microbes will immediately perish in the stomach of the Anopheles mosquito. The only survivor is the malaria protozoon.

BILHARZIASIS

This is another tripartite parasitic arrangement. An infected peasant passes the eggs of the parasite in the urine or feces. This usually takes place in the open fields, and usually at a riverside, in order that he may wash as well. The eggs immediately hatch, and little worms called meracidia rush all around in search of a certain fresh-water snail. If they don't find it within 48 hours they die. The one which succeeds in locating the snail pierces its soft body cover, and immediately reaches for the liver, settles down, and multiplies into hundreds and thousands of another form of little worms called cercaria. The snail soon dies because of its crowded liver; thousands of cercaria get released into the fresh water again. This time they are looking for their original host: Man. If they don't reach him within 48 hours they die. They find him washing himself in the water, or toiling bare-foot in the wet fields, earning his meals. The cercarias pierce the human skin, causing a bit of itching. The patient thoroughly scratches the site, further pushing them inside the skin, until they reach the small blood vessels of the skin. From that moment on, it is business as usual.

The parasites know very well the map of their host. They navigate very accurately into the liver, the great food store. There, they enjoy a blissful existence, and grow into males and females, who ultimately start reproducing. When the pregnant female wants to lay her eggs she reaches for her map again, and starts navigating *against the stream*, in the portal veins, which collect blood from the gut and bladder.

The females of the bilharzial variety called Bilharzia Mansoni go into the walls of the colon and rectum where they lay their eggs. These eggs are destined to go down with the feces. Each egg possesses a side-pointing sharp spike which tears the mucosa with the contractions of the colon muscles, in order to reach the lumen, to join with the feces.

The females of the other variety called Bilharzia Hematobium share the journey down to a certain point, then they fare good bye to the Mansoni friends. Their new route takes them to the walls of the urethra and urinary bladder, where they lay their valuable eggs. These are destined to pass out with urine. Each egg possesses an end-pointing spike, more suited to the bladder muscle contractions. It pierces the mucosa, and ends swimming in the urine.

The perfect knowledge imprinted in the genes of the Bilharzia worms includes the following *strategic items:*

(a) Fully updated maps of the human anatomy, as well as psychological study of the social habits of millions of peasants in agricultural country life. (b) Knowledge of the osmotic characters of the urine and the river water. It adjusts its eggs to an optimum which allows them to hatch only once they touch the fresh water, and not before. (c) Strategically placed spikes in the eggs, suiting either the colon or bladder walls accordingly. (d) Skin irritating chemicals which force the newly infected boy or girl to scratch hard, to guarantee pushing the cercaria into the skin blood vessels. ***Evolutionist reasoning says that all this planning and imprinting had been executed by purely random chance.***

THE EUROPEAN SONGBIRD

A certain European songbird called Thrush is commonly infected with a fluke worm which lives in its gut. Its eggs get passed in the bird's droppings, which may get eaten by a certain grazing snail. They hatch inside the body of the snail and multiply in its liver. When it gets ready to infect another bird again it performs a really funny trick. It migrates into the tentacle of the snail, stretching it into a pulsating transparent bulge. This bulge looks very much like a wriggling little worm, perched on the head of the snail. The infected snail also gets sick and behaves very oddly. Instead of hiding under the leaves during daylight, it stays on the top of the leaf, almost advertising its cargo to passing birds. Soon another songbird gets attracted to the scene. Believing that the snail's tentacle is a juicy worm it picks it off. At last the larva *has lured* another victim, and has reached the intestines of a new host.

Figure (10) the larva on the head of the snail trying to lure a bird.

One evening I enjoyed watching this drama on one of the famous video films of nature made by David Attenborough. At the climax of the film the bird kept flying off and then returning back, looking at the worm-like tentacle from all sides. When it finally picked it off there were several aahs! And oohs! by the audience. But the most significant of these were the words *"OH MY GOD"*, unwittingly uttered by two devout atheist, very dear, friends who attended the party.

EXAMPLES OF SYMBIOSIS

The literal meaning of symbiosis is: *living with, and depending on, each other.* There are thousands of examples in natural life. I have chosen only two glaring examples, which vividly illustrate this genetically imprinted co-ordination between totally different species: Acacia trees, and ruminating animals.

ACACIA TREES

Two hundred million years ago the Queen of a certain tribe of ants signed a pact with the first seeds of Acacia trees. In this pact the Acacia tree agreed to grant three favors to the tribe of ants. The first favor is allowing the ants to dig cozy shelters inside the trunk, as well as the bigger branches. The second favor is supplying a special sugary food for the adult ants. This gets secreted in the form of balls of honey on the surface of the acacia leaves. The third favor is also in the catering department, but this time as a special food for the children ants, and not the adults. This gets secreted as small balls of protein matter which the adult ants pick off the leaves, and take inside the tree, to feed the newly hatched larvae.

As you see, the Acacia tree, which belongs to the plant kingdom, has full knowledge of the details of the social habits of the colony of ants. In return for food and shelter, the ants agreed to act as body guards, and to offer protection to the Acacia tree in several ways:

At the slightest shake of a leafy branch, by a giraffe, a monkey or a leaf-eating insect, thousands of fighter ants rush into the scene. They keep stinging and attacking the trespasser until he is forced to leave the tree in peace. Whenever a climbing plant tries to climb up the stem, of their host tree, thousands of ants attack the climbing tendrils while they are still small, and virtually tear them to pieces. The ants also patrol an area of twenty meters diameter around their tree friend. They specifically devour the tiny saplings of any new tree in that protection zone. This will prevent any other tree from future overcrowding our tree, not only in the food sources at the roots level, but also in the sunshine at the tree tops. This is a sophisticated built-in genetic long-term strategy in both ants and acacia trees.

RUMINATING ANIMALS

This group includes many herbivores, such as sheep, cattle, goats, deer, and camels. In the wild they graze as quickly as possible, and then rush into a safe area away from lions and tigers. There, they ruminate, by regurgitating the grass back into their mouth for proper mastication, at ease.

The main bulk of their food is cellulose, which their stomachs cannot digest. Zoologists have discovered that a certain bacteria do help them to digest that cellulose. Without the symbiotic assistance of this bacterium these animals would die of starvation. This bacterium is not present in the stomach of the new born calf. The animal receives its first quota of this helpful bacterium with its first grass meal in the wild.

Imagine the millions and millions of herds of these herbivores all over the Earth, being totally dependent on such an alien tiny micro-organism for their very existence. Just imagine the calamitous mass-extinction that would take place if such a micro-organism would disappear from the surface of the Earth. It will include not only these herbivores, but also many millions of carnivore species as well. It could even bring the whole life on Planet Earth into a complete halt.

EXAMPLES OF SYMBIOTIC POLLINATION

I'll quote here only two of interesting examples: the Hammer Orchid, and the Rafflesia Flower.

THE HAMMER ORCHID

Let me start with this beautifully colored example, which has also got an appealing smell. The flower called Hammer Orchid has a funny habit of manufacturing its lower petal not in plant shape, but in a shape which belongs to the animal kingdom. It resembles the belly and the hind end of a certain female wasp. Not only that, but it also gives away the same *scent* which emanates from the body of that particular female wasp.

The appealing smell attracts the males of that particular wasp from all over. A male would gently, and eagerly, land over the female-shaped petal. In a few seconds it decides that it is a fake-shaped female, and off it goes. The interesting thing in this farce is that when the male flies off it takes on its head thousands of the pollen grains of the orchid flower. The pollen-carrying stamens are strategically positioned ahead of the female-shaped petal, and the male wasp cannot avoid picking them up before leaving the flower. It similarly visits hundreds of orchids everyday, dutifully gathering and spreading its valuable cargo of pollen.

As we mentioned before, research techniques have greatly improved in the last decade. It was clearly documented that no other insect or bird is able, nor even tries, to pollinate this particular orchid flower. If that particular wasp disappears from the surface of Earth, so will this particular orchid flower. How could this exclusive relationship take place? This is *NOT* Natural Selection. Take a seed from this particular plant, and an egg from this particular wasp, and place them side by side under your microscope. One belongs to the *plant kingdom*, the other to the *animal kingdom*. The set of genes in the orchid seed has got complete knowledge of the habits and preferences of the animal form of life which is manufactured by the wasp egg, up to the exclusive perfume it dons during sex!

In order to get this exclusive and detailed degree of genetically imprinted coordination between two forms of life there are only two possibilities. The first is as follows: The originators of each of the two species may have experimented a lot, and then got together over a cup of coffee 200 million years ago, and laid together the foundations of their strategic coordination. They signed the documents, and then implanted the necessary instructions into the genes of their descendants!

The second way is the logical one: the genes for such coordination between two genetically separate species have been an integral part of the whole drama of life:

> It must have been predetermined since eternity in the original
> seeds of life. Those seeds have evolved into the present delicate
> balance between animal and plant life on one side, and on the
> other side into the balance between parasites and their hosts.
> These are the two major balances that characterize life on Planet
> Earth. The gene-Suppression-Activation-Theory is presented a
> few pages later. It gives a scientifically feasible explanation of
> how such balances could have followed a premeditated master-
> plan. This is the Principle of Deism.

THE RAFFLESIA FLOWER

Rafflesia is a type of fungus which makes the largest flower on Earth. It is more than three feet in diameter, too heavy for it to have a stem, and so it rests on the ground. It is beautifully colored, but its smell is horrible. It emanates the odor of stinking meat, hence its name *"The stinking corpse Lilly"* The stamens carrying the pollen grains are hidden inside a special corridor inside the flower. The horribly foul smell attracts flies, the specialists of foul odors! Swarms of them come to the flower, searching for the source of the odor. When enough of flies have entered the dark corridor its outlet gets closed for 12 hours. It opens again in the morning, releasing the flies with their cargo of pollen, stuck to their wings and bodies. The flies fall into the same trap in another flower, dutifully transporting the pollen everywhere. Zoologists have found that the stigma, which is the female part of this flower, matures long enough after the stamen, which is its pollen-forming male part, to guarantee cross fertilization from another flower, rather than incest, (self), pollination. No bird or bee or butterfly can dare to get inside to get that pollen, the foul odor makes it an exclusive game reserved for the flies.

A WORD OF INSIGHT

There are thousands of similarly exclusive relationships between various plant species and their pollinators. The wisdom and purposefulness is evident: to guarantee that collected pollen would reach its ideal destination, which is a flower of the *same* kind, and not just any flower. It is as if each pollinator has become an exclusive special-delivery courier employed by each particular plant.

Simultaneous concerted trial and error experiments, carried out by each plant and its animal pollinator are nearer to hallucination than theory or speculation. It is also ridiculous to try to explain it away by haphazard-chance.

AN EXAMPLE OF SYMBIOTIC TRANSPORT, THE MISTLETOE PLANT

This is a plant which can only grow as a parasite, and can never grow on its own. It gets its lifeline by pushing its root into the juicy branches of other trees. In this pampered situation it grows and forms flowers and seeds. Its seed fails to germinate if it gets dropped onto the ground. Even if it sprouts on the ground the roots will fail to collect enough food and water and the sapling will die. The Mistletoe has got to invent guaranteed means of transportation, from the branches of one host tree into another. The pact which it made with a certain bird is sure-fire, *down to the smallest small-detail.* This bird feeds exclusively on the seeds of Mistletoe, and is called the Mistletoe Bird. Its beak is *tailor-made* to squeeze the seed out of its sac. The seed gets swallowed whole, soaked in a sticky nutrient gel. The bird likes that gel, its gets digested in its gut. What is passed out in the dropping is only the sticky material in it, strongly adherent to the undigested seed. The bird dropping *does not fall on the ground*; it stays firmly hanging from the anal opening of the bird due to its sticky character. The bird gets rid of it in a funny side-dancing movement. It uses one foot to attach the dropping to the tree branch, and then moves sideward cutting off the thread-like connection to its anal orifice. Then it flies away, leaving the seed ***securely glued*** to the tree branch. Within 24 hours it germinates, pushing roots into the stem of its new host. Within days a new Mistletoe parasite plant is firmly established, and will happily live ever-after.

Figure (11)

The Hammer Orchid flower. The seeds of this plant have complete knowledge of a certain member of the animal kingdom, including its sexual habits and preferences, even the perfume it dons during sex.

The Rafflesia Flower, beauty that stinks.

EXAMPLES OF CAMOUFLAGE AND MIMICRY

There are thousands of examples of land and marine life forms which change their colors to suit their surroundings. Sometimes the color changes are variable from one minute to another, which suggests that they are a reflex performed through vision. The examples given here are totally different: a snake-shaped caterpillar, and the *international-poison-sign*.

Look into the tail-end of this harmless 4 centimeter long caterpillar, in figure 12. It is a perfect replica of the head of a dangerous snake. Modern continuous photography over several hours and days has proved that this similarity does scare the birds away, and allows the poor caterpillar to live in peace. First, how did this caterpillar *know* that birds are scared away by snakes? Second, how did it decide to take advantage of that knowledge, and get it coded into its genes, in order to get its tail-end painted that way?

Lyall Watson gives an interesting comment about this worm by saying: *"This meaningful and inventive mimicry requires the existence of an influence which goes beyond the bounds or the abilities of the individual, or the species."* [Watson, L., 1995, page 64]

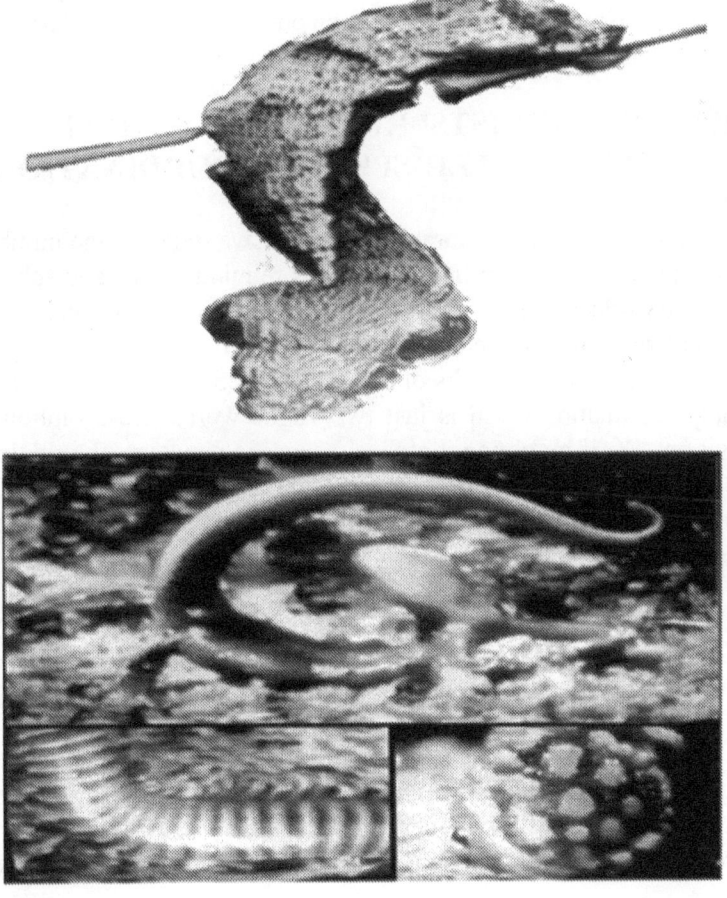

Figure (12), explained in text.

The beauty of the *international-poison-sign* is that it is literally international. It is the combination of black and orange colors, (in the form of spots or stripes), on the skin of an insect, worm, or animal. It means that this insect, worm or animal is passively poisonous. If it is eaten the predator may lose its life, or get seriously hurt. This picture shows three of these examples: the salamander, the poisonous frog, and the famous centipede. Snakes, monkeys, birds, and all sorts of predators avoid any prey advertising this sign. This sign has gone into the instinctive genetic information of all predators. This fact, alone in itself, proclaims an organizing committee, beyond the abilities of both predator and prey. The most spectacular part is even yet to come: some species *fake* this sign, and use it as protection. A moth, called the *clear-wing-moth*, is not poisonous, but its genes do manufacture the fake sign on its hind part, to get protection from the birds, free of charge!

THE EVOLUTIONISTS' RESPONSE TO THESE EXAMPLES OF INTER-SPECIES COORDINATION

The evolutionists' library is as extensive today as the number of books defending various religions, various religious sects, and sub-sects, all over the world. Get any book on evolution, and look into the pages which talk about such highly exclusive symbiotic pollination, or the subject of parasitism, or the funny tricks of camouflage and mimicry. You will find the only explanation given is just two vague words: Co-evolution, and adaptation. As if by giving it a quasi-scientific new term the riddle gets solved. In whole books or encyclopedias you find no more than a few lines or words to explain away how such masterpieces of coordination could come to exist.

"Encyclopedia Americana", [1982, vol. 21, page 288] says: *"By accident, or experiment they, (the parasites), have discovered the advantages of living on the expense of their fellow creatures."*

The words *"by accident"* could mean the oft-repeated chance-mutations. But I find the expressions of *"experimenting"* or *"discovering"* to be a very odd scientific comment. In the 1994 and later issues of the same Encyclopedia, prudently enough, these odd phrases have been omitted.

"Guinness Encyclopedia of the Living World ", 1992, page 17 says: *"The whole evolution of flowers and the animals, especially insects, which pollinate them, provides astonishingly specialized examples of*

Co-Evolution." These were the only words of explanation given by that encyclopedia. I found the same explanation also even in 400 or 700 paged books totally dedicated to the subject of Evolution.

THE FUNNIEST WAY EVER TO END A BOOK!

Please let me here quote for you the last page, which summarizes the wisdom of a whole book titled *"Does God play dice?"* [Stewart, 1997] He has concluded that all the complimentary aspects of life on Earth have just evolved out of chaos, with no guidelines at all. He ends the book by saying:

" ... A real world of complicity is the evolution of bloodsucking. This occurred when the rules for the early mammalian anatomy (blood) interacted with the rules of an ancestor of the mosquito (which has developed an organ for sucking liquids, probably water). By a creative coincidence, the water-sucking proboscis happened to be able to penetrate human skin. This collision of two developmental spaces caused them to co-evolve in a new way, not inherent in either developmental space on its own. The result of this complicit co-evolution is an insect that is adapted to sucking human blood. One consequence happens to be malaria, a more predictable simplicity erected on top of the initial complicity ..."

That was in the final page of a supposedly scientific book.

FEASIBILITY OF CHANCE MUTATIONS

Chance-mutations, by random throws of dice could never give a logical answer. There is so much at stake during information gathering, then experimentation, and feed-back, and implementation, etc., in order to perfectly implant the rules of such coordination *into the genes* of two or three *separate* species. According to the rules of probabilities it would take many more billions of years than the age of the whole Universe.

At the long end of every such discussion, whether in a lecture or in a book, and without a single exception, you will get a casual hand wave. Whichever evolutionist is lecturing, he will magnanimously refer you to no less than Charles Darwin himself by saying: *"This subject has been extensively dealt with in Darwin's book titled "The Various Contrivances by Which Orchids Are Fertilized By Insects"*, please let us proceed to more important items." When ultimately you get to that out of print book you will find that Darwin has cleverly described most of the tricks of orchid pollination. His observant eye has even predicted the existence of a certain butterfly with a 25 cm. long coiled suction tube, when he observed a flower with that deep sac of nectar in an African country. Later, scientists actually discovered that butterfly. Well and good, but the only explanation given in the book in the year 1882 by Mr. Darwin was nothing more but the same term of *Co-evolution,* with no further explanation at all!!

CHAPTER 7

* * * * * * * * *

G - SAT

The Gene-Suppression-Activation-Theory

* * * * * * * * *

"You have to come up with an idea, work it through, and only then figure out how to test it experimentally. In that sense, science is an art."

Albert Einstein, April 1950

The G-SAT Theory I am proposing in this chapter, is as scientifically logical as Einstein's Theory when proposed in 1915, before the famous experiment of May 1919 that validated it. Similarly, with the recent explosion of gene-science, it looks quite feasible to have the G-SAT theory experimentally tested and proved. The proof is both scientifically and practically feasible. It could be around the corner in less than a decade.

Dr. Hussein A. Amin, 2005

Today, the scientific community can safely presume that if a scientist had lived one second after the Big Bang he could have accurately predicted what happened later. Armed with the knowledge of the amount and type of *matter* around him, and the *physical laws* governing it, (and hopefully having a super-computer!), he could have calculated, in advance, everything that took place until the formation of our solar system and earth. He wouldn't have bothered to ask what happened one second earlier at the actual moment of the Big Bang, because there would be *no way* to know.

Similarly, I can safely presume that if a biologist would have co-existed with the first single-celled organism, 3,500 million years ago, he would have done the same. Armed with the knowledge of the allegedly junk DNA, as well as how the genetic bases, (Adenine, Thymine, Cytosine, Guanine and Uracil) do their work, he could have accurately predicted how *life* on planet Earth would progress. He wouldn't have bothered to ask how this first single-celled organism came to be, because there would be *no way* to know.

*Biologists routinely regard Junk-DNA as insignificant. On the contrary it is the archive of past **and future** life, encoded since eternity, on about 97 % of the DNA strands of every single cell in every form of life, including you and me.*

JUNK-DNA

It has become an established fact that each fertilized egg contains many more DNA than is needed to create a whole adult body. These are variously called surplus genes, introns, and Junk-DNA. Junk-DNA forms more than 90 % of the total genome of every form of life. Nobody knows what they do, or where they came from. Junk-DNA is present in the cells of every multi-cellular and unicellular organism. These inactive genes get carried over from the embryonic cell into every body cell, and ultimately into the sex cells of the adult, and from them into every next generation, in all forms of life. The funniest explanation of these inactive genes was a suggestion by evolutionists that they are just parasites, carried over from one generation to another, with no purpose at all. *I believe this Junk-DNA IS the archive in which all the past **and future** history of life is preserved.*

Whenever a new species is created, the genes of its ancestor get suppressed and become a part of what we call junk-DNA, relating to the past history of the newly created species. On the contrary, the junk-DNA of the grand-ancestral species, the blue-green algae, and later bacteria, were like a storage archive for *future* species formation. In that case of *future archives,* the activation of one dormant gene could initiate a whole set of daughter genes. This would explain the discrepancy in the numbers of genes. They were and still are, only 3,000 in the blue-green algae, the grand ancestors of all life. Now the number of genes is around 30,000 in all warm blooded animal life, including Mankind.

Modern animals did not evolve from each other. Hence, Mankind did not evolve from monkeys. This is simply a layman's incorrect inference. Instead, any two modern animals have a common ancestor at some ancient time in the long history of life. The archive of all life-ancestries is still present in the genes of all modern animals. It occupies the silent segments of the genome, which is wrongly called *junk*-DNA.

What initiates the suppression of certain genes, and the activation of others, in order to create new species? I believe this is affected by environmental physical factors, *according to a pre-determined Master-Plan.*

THE TEMPORARILY SUPPRESSED BUTTERFLY

I wish to illustrate for you the magnitude of Gene-Suppression, and what it can do. Look at these two beautifully colored pictures, (fig. 13). The first is an everyday butterfly which we see jumping around in our gardens, sucking nectar, and spreading pollen from one flower to another. The second is a caterpillar, which lives on plant leaves, chewing them up. They are two aspects of the same creature. Every cell in each of them contains the same set of genes contained in the other. At one stage of its life, a group of these genes are activated, with suppression of all the rest, we get the caterpillar. At another stage of its life, the first group of genes gets suppressed, with activation of all the rest and we get the brightly colored butterfly.

SOCIAL INSECTS, WHICH GET
SUPPRESSED IN ALTERNATING WAYS

There are several 'castes' in each nest and colony of ants, termites and other social insects. The 'castes' include queens, small-sized workers, medium-sized workers, soldiers, honey-pot ants, etc. The differences in body size, body shape, and functions vary greatly. But each of them has got the same set of genes. Different genes are switched on or off under different rearing conditions, different diets, or environmental factors. In termite colonies, when the egg-laying queen dies, one of the female workers will simply metamorphose. A certain set of its genes become switched on. The switch-on is apparently caused by the disappearance of certain pheromones previously secreted by the deceased queen. The body form of that female worker will change, and it will simply become another queen.

THE ETERNALLY SUPPRESSED CHIMPANZEE

We already know that there is only a 1 percent difference between the genes of Mankind and those of chimpanzee or gorilla. When you compare these two pictures of a man and a chimpanzee you recognize less difference, than that which exists between the caterpillar and butterfly in the other picture, (figure 13). As we said before, the latter two are one and the same, but with many alternately suppressed genes in each. The cells of both Mankind and gorilla also contain many inactive genes, the same as all living organisms.

This is the theory which explains the mystery of HOW Evolution does its work: Evolution creates new species by a masterly handling of the 97 % Junk DNA. This Junk-DNA is present in every cell since the start of cellular life, 3,500 million years ago.

A new species gets created whenever certain segments of this surplus DNA get activated, or, when certain active segments get suppressed. Either way, this process of species creation has been going on since eternity, according to pre-determined master-plan. The characters of the new species get fine-tuned over the years by the mechanism of Natural Selection, in order to fit better into different habitats.

Species creation does *not* take place by haphazard chance-mutations (and then get sifted by Natural Selection). We have discussed the pure random chance probabilities of creating even the simplest first step of the chlorophyll or hemoglobin molecules. We have found that species-creation in this way would take billions and billions of years longer than the life time of the whole Universe.

Mankind was able to breed selective characteristics in the species of dogs by using guided artificial selection. He could do all that in a mere few thousand years. If dogs were left to themselves we would have never ever got the tiny playful Chihuahua. Nor could the police have got the dogs with greatly augmented sense of smell, which are able to sniff a trace of heroin in one bag out of thousands of luggage in an airport. This is what guided artificial selection, or breeding, can do in a few thousand years time. It is also comparable to what guided natural selection, according to our theory, can also do within the time limit of life on Earth.

This rationale automatically invites an important question: What are the physical factors that could initiate various steps of species-creation? In many animal species there are observable biological cycles, which are governed by 'genetically regulated clocks'. The most common examples are the mass migrations of fish and birds. The mother bird does not tell the young ones where to go. So it has to be inherited, which means it is encoded in the genes. Guided by, still unknown, physical means, the genes in the bird's little brain do tell it where to migrate. These genetic clocks are fine-tuned by external physical and atmospheric influences. Herewith we will describe a highly significant example, significant from the point of view of the Gene-Suppression-Activation-Theory of Creation. It is the story of the Christmas Island crabs.

Figure (13)
Each butterfly and caterpillar is the *SAME* genetically, with two sets of genes alternately suppressed. This shows the magnitude of what gene-suppression can do. Their bodily differences far exceed the differences between this Chimpanzee and this Man. Only I % genetic difference has kept this Chimpanzee poking for termites since millions of years ... While the man on the left, has started exploring his Universe, and is telling his grandson about it.

CHRISTMAS ISLAND CRABS,
A MIRACLE OF TIMING

These crabs live on Christmas Island, in the Indian Ocean 200 miles south of Java. Their population is estimated at 120 millions and they are found nowhere else in the world. An adult's crab size is about five inches and they live in burrows on the forest floor.

The timing of breeding of Christmas Island crabs takes place in a most extraordinary way. The millions of adult female crabs have got to lay their eggs on the beach, on a single specific night out of the 365 nights of the year. Not only that, but the hour is also specific, it is at midnight. If they miss that night and that hour, then they will lose the chance to produce a second generation for a whole year. The critical night is a one in November when the moon is in its third quarter, and the tide is rising. Those crabs are land animals; they breathe air and cannot swim. If they fall into the ocean water their heavy shells will drown them. *But*, their eggs cannot hatch *unless* directly deposited into sea water. These two contradictions have been solved by a most ingenious trick of Nature.

One month earlier the adult male and female crabs suddenly emerge from their forest floor burrows, and start marching in millions towards the beach. They have been doing this for millions of years. During that procession nothing deters them or changes the direction of their march, even if thousands get crushed by cars while crossing modern traffic roads on the island. On reaching the beach the males excavate sand burrows where they mate with the females, and then march back to the forest. The females stay in these burrows for two weeks while the eggs mature. They get ready for spawning just in time for the rising tide on the specific night. At the height of the tide, the beach gets covered by shallow water into which the millions of pregnant females rush in a great hurry to lay their eggs, before the water recedes. They are still able to hold to the ground in the shallow tide, so that they won't get carried into the ocean. Once every one of them lays her eggs she raises her claws above her head in a saluting gesture, before she starts marching back to the forest. Each female lays about 100,000 eggs which get swept to the ocean with the ebbing tide. They hatch and become a part of the big amount of zoo-plankton in the sea. They live by eating smaller plankton, and they get eaten by bigger plankton and by fish.

Figure (14)
Female Christmas Island crabs depositing their eggs in sea water

Once every five years the high tide of December brings thousands of tiny crabs back to the same beach. The reason for this five year cycle still defies our knowledge. The returning crabs are tiny, all of sizes no bigger than little ants. They are very light and stay on the surface of the water. They are the only survivors of the millions of eggs laid on the same beach one month earlier, five years ago. They valiantly march across the whole island to join, at long last, their ancestors on the forest floor. The odds against their survival are huge, but they overcome that by the sheer numbers spawned each year. An adult female lives for ten years, during which it lays a million eggs. If only two of them survive and come back to the island then the population of Christmas Island crabs is safely maintained.

THE TIMING AND THE MYSTERY

Now let us ponder a little about the mysterious timing of this egg-laying. It is surely achieved through instinctive instructions laid down in their genes. There is no way that their ancestors could have gone through *trial-and-error* experiments until they could accomplish this well-timed feat. Nor could all these coincidences fit together by pure random chance.

Modern science has explained much odd animal behavior, which were just mysteries before. We now know that flocks of migrating birds are guided, at least partially, by the north-south magnetic field of the Earth during their long range flights. Thousands of salmon fish return to lay eggs in the same rivers where they have been born. We now know that this is guided in part, by certain faint smells of those rivers, which get imprinted on their little brains once they hatch. It is amazing how they keep remembering that smell, and how they follow its faintest traces across the oceans of the world. Grazing animals in Africa seem to collectively recognize certain signs of an oncoming drought. This gets translated into an almost sudden collective migration, into a new habitat.

All these behaviors are surely genetically imprinted. All the animals do is obey their orders, orders which haven't changed for millions and millions of years. The Christmas Island crabs, happily enjoying their life on the forest floor, get an important message in October every year. Perhaps it is a certain degree of humidity, a certain difference between the length of daylight and night darkness, or a certain change of the magnetic field caused by the moon. Once they get whatever that message is, everyone leaves what it is doing, and off they go for a yearly rendezvous at a certain specific spot and a certain specific night and hour on the beach.

The physical and environmental factors involved in these remarkable feats of timing may not be clear to us now. But these examples clearly show the magnitude of their effect on the life cycles of all forms of life.

THE SCENARIO OF LIFE, UNDER THIS THEORY

Now let us go 3,000 million years back in the history of the Earth. The blue-green algae and bacteria are unicellular organisms. Both were quite abundant, and were filling the whole Earth since that time until today. Each has a genetic make-up of approximately 3,000 genes, and a lot more of superfluous or inactive DNA. Blue-green algae were the first form of cellular life on Earth; they monopolized the Earth for almost 2,000 million years. They contain chlorophyll pigment which manufactures oxygen, as well as another blue pigment which partially masks the green colour of chlorophyll, hence their name.

"Much later on in the life of the Earth there appeared the green algae, which are still present today. Their color is grass-green due to the presence of unmasked chlorophyll. This is also the characteristic of higher green plants, such as mosses, ferns, and flowering plants. This fact supports the hypothesis that the higher plants have evolved from the green algae." *[By courtesy of "Encyclopedia Americana", 1994, vol. 1, page 552, under title 'Algae'.]*

"Bacteria have to have been widespread on Earth at least since the middle Proterozoic Era, about 1.5 billion years ago, *when oxygen appeared in the atmosphere;* the cyanobacteria were responsible for this remarkable global occurrence." *[By courtesy of "Encyclopedia Britannica", 1995, vol. 14, page 584, under title 'Evolution of Bacteria'.](Emphasis added).*

I'll summarize another quotation from the latest Britannica, 2002, vol. 14, page 1156 in my own words as follows: [The Cambrian-explosion started about 540 million years ago. Most of the phyla (classes) that characterize modern animal life started to appear at that time, they evolved over a period of only 5 to 10 million years. In evolutionary terms this number of years is almost a flash of suddenness. The reasons for this Cambrian explosion are still debated, but a leading theory is that the amount of oxygen in the atmosphere has finally reached levels that allow large complex animals to exist. Oxygen levels may also have facilitated the metabolic processes that produce collagen, a protein building block that is the basis for hard structures in the body.]

On page 1157 the Britannica 2002 says: "Terrestrial plants are believed to have evolved from the chlorophytes, such as the green algae."

These three quotations from the two major Encyclopedias look so much like the jagged pieces of a jigsaw puzzle, which beautifully fit to illustrate a magnificent view: The blue-green algae spent 2,000 million years as the sole representative of life on Earth, all the time busy filling the atmosphere with oxygen. At some time, at a certain level of oxygen, some of them started to *suppress* some of their genes, and the blue pigment disappeared. The result was the appearance of green algae, grass green in color. The green algae also contributed to the overall build-up of oxygen, and became the ancestors of all plant life forms. They contained enough suppressed genes in store, to account for the *FUTURE* unfolding of all the later events that took place in the history of evolution of the Plant Kingdom. Over thousands or millions of years, more and more of the suppressed genes in the cells of these green algae became activated. Their activation followed a program which had been pre-determined since eternity. Specific activations coincided with specific physical habitats, heat, humidity, salinity, rains, droughts, etc. Whenever a genetic activation took place, a new species of plants emerged, until they filled the whole Earth. Herbs, grasses, flowers, ferns, giant sequoia trees, *are all descendants of the grass-green algae.* It all happened by a sequential shutting-off and activation of various genes to fit into the various ecological niches throughout life on Earth.

OXYGEN LEVEL AS *ONE* FACTOR IN INITIATING THE *ANIMAL* KINGDOM

About 1,500 million years ago, at another certain level of oxygen, some algae had some of their suppressed genes activated, and other genes suppressed. This created other uni-cellular organisms called bacteria, which did not have chlorophyll anymore. They also had cell-membranes made of proteins rather than cellulose, and they could directly use the atmospheric oxygen for their metabolism and multiplication.

This oxygen-consuming species is the grand ancestor of all animal life forms. It contained enough suppressed genes in store to account for the *FUTURE* unfolding of all the later events that took place in the history of evolution of the Animal Kingdom. In a way exactly similar to what took place in the Plant Kingdom.

SOME OTHER POSSIBLE PHYSICAL FACTORS

The oxygen level is only one single physical factor that could determine the cycles of suppression-activation of different genes. The stories of Christmas Island crabs and of mass migrating animals also illustrate thousands of other physical factors that could have contributed to the cycles of gene-suppression-activation.

'*Scientific American*' reports in the March 2001 issue on its page (79) about the ions of sodium and potassium. Potassium exists in the cells of both animal and plant kingdoms. But sodium exists outside the cells in the animals and is very rare in plants. Hence herbivore animals will diligently search for sodium sources outside their purely plant foods. Sodium exists as one percent of sea water by weight, but also exists as solid material in many soils, rocks and caves. Potassium in sea water is a minimal trace. Its isotope, (potassium 40), is radioactive. Its decay-rate is used as a timing clock for the ages of the sun, earth, rocks, and fossils.

The 'Scientific American' article gives an explanation for the sudden appearance of multi-cellular life at the time of the Cambrian explosion, 540 million years ago. It states: **"Perhaps multi-cellular life waited for the decline of potassium's intrinsic radioactivity. It is possible that radio-active potassium has always imposed some waiting period between the birth of a new sun and the rise of complex life on its planets."**

In a similar vein, the *Ozone protective blanket* mentioned on page 37 had to be formed first; before complex life forms could start to appear.

THE G - SAT THEORY AND MANKIND

The unfolding of the plant and animal kingdoms was orchestrated by the carefully balanced activation of previously silent genes in ancestor species. The cycles of activation of suppressed genes, and vice-versa suppression of previously active ones, were initiated in each case by myriads of physical factors. Some of these physical factors are recognized by our senses, like sunshine, heat, cold, humidity, oxygen and carbon dioxide levels, etc. Other factors are recognizable by our modern scientific equipment, such as magnetism, gravity, solar winds, cosmic radiations, etc. But still there are hundreds of other physical factors, single or combined, that we hardly recognize at all. With every step of these changes, the newly evolved organism carries in its genetic make-up all the suppressed genes of its predecessors, since eternity. No wonder that the cells of gorilla or mankind have this load of ninety seven percent of *Junk-DNA*.

What could have been the factors which suddenly activated the genes of nakedness, language, and inventiveness, about 30,000 years ago, thus creating Mankind? Was it written since eternity that at a certain level of balance on earth this momentous occurrence would take place?

SCIENTIFIC AMERICAN Magazine suggests: "that a brain mutation roughly 50,000 years ago had the lucky effect of rewiring the human brain and thereby unleashed the ability of symbolic thoughts, language and innovation." [June 2005 issue, pages 66 – 67].

The differences between a caterpillar and its butterfly are much greater than the differences between blue-green-algae and bacteria, or between the chimpanzee and Mankind. As we mentioned a few pages ago, both caterpillar and butterfly is one and the same individual, having the same set of genes.

At this point it is very interesting to remember that almost 99 percent of the genes of Mankind, gorillas, and chimpanzees are exactly similar. This is common knowledge now. The junk-DNA is similar in all three. The one percent difference is in the *active* part of DNA. This 1 percent difference is what causes the gorilla to be warm and hairy, as well as much more muscular than we are. It makes their arms, hands, feet, and face different from ours. But most important of all, this mere one percent difference is what makes us able to *invent, and talk.* If only we could somehow get the junk-DNA in a human cell activated, it would unfold a magnificent video-film, which would betray the whole story of *guided natural selection*, and the creation of plant and animal life on Planet Earth.

If only we knew which segments of the chromosomes to change position, or divide, or delete or snip, or clip, we could become GODS! We could even create some forms of life which became extinct millions of years ago. Just imagine if scientists could clone a few baby dinosaurs out of a cell from your skin. Ten years ago, this mere idea in itself could have given writers of science fiction a limitless supply of stories. But today, after cloning of sheep and cats, the idea looks quite feasible, and even much more exciting.

'Nature' magazine reports, (*Jan. 16/2003*), that a certain stick insect has lost its wings 50 million years ago. But recently a new variant of that insect has regained its wings and uses them for flight. The new wings did not re-evolve from scratch; genetic blueprints seem to have lain in wait for at least 50,000,000 years. They were re-activated when regaining the flight ability became more favorable in a changing habitat.

HUMAN EMBRYOLOGICAL EVIDENCE

If you are a physician go back to your embryology books. You will remember that the early human embryo does form fish-like gills, which soon degenerate, after which he forms the proper human type of breathing organ, the lungs. At its tail end, the early human embryo forms an actual sizeable monkey-like tail for some days, after which it also degenerates.

Even more significant is the way he develops his kidneys, this is done in three different stages, the pronephros, the mesonephros, and the metanephros. The pronephros is formed in the first three weeks of life, high up near the heart. It is a series of six to ten pairs of tubules, very much like the kidneys of primitive aquatic vertebrates and eels. They disappear by the fourth week of fetal life, leaving a trace called the Wollfian duct. The next stage is the mesonephros which forms in the second month of fetal life, around the middle of the body. It greatly resembles the kidney organs in primitive land vertebrates. This also disappears during the third month of fetal life. Sometimes it leaves traces of tubules or ducts which may cause trouble in adult life. The third prototype, called metanephros, is *the final blueprint* of renal development. It starts to develop by the third month, nearer to the hind end of the body. The new kidney grows upwards, forming our familiar pair of permanent kidneys.

The first and second blueprints are initiated by genes which make kidneys in fish, eels, and primitive vertebrates. They soon get *suppressed* and replaced by the *activation* of other genes, which take over to give us a more advanced category of kidneys for the rest of our lives. The two suppressed blueprints do not disappear but remain silent in every one of your cells and mine. Future generations will repeat the same cycles again and again, for eternity.

This is a glaring practical example of the Gene Suppression Activation Theory, as it is presented in this book. Thousands of physicians have studied embryology, and know all these details by heart. But it has never occurred to us before, to notice their hidden significance.

Here are genes that try to remember their past, and attempt to manufacture fish-like kidneys. Soon they get new orders to shut-up, and never to function again. They do not evaporate; they remain silent in every one of our cells, for us to call: Junk DNA.

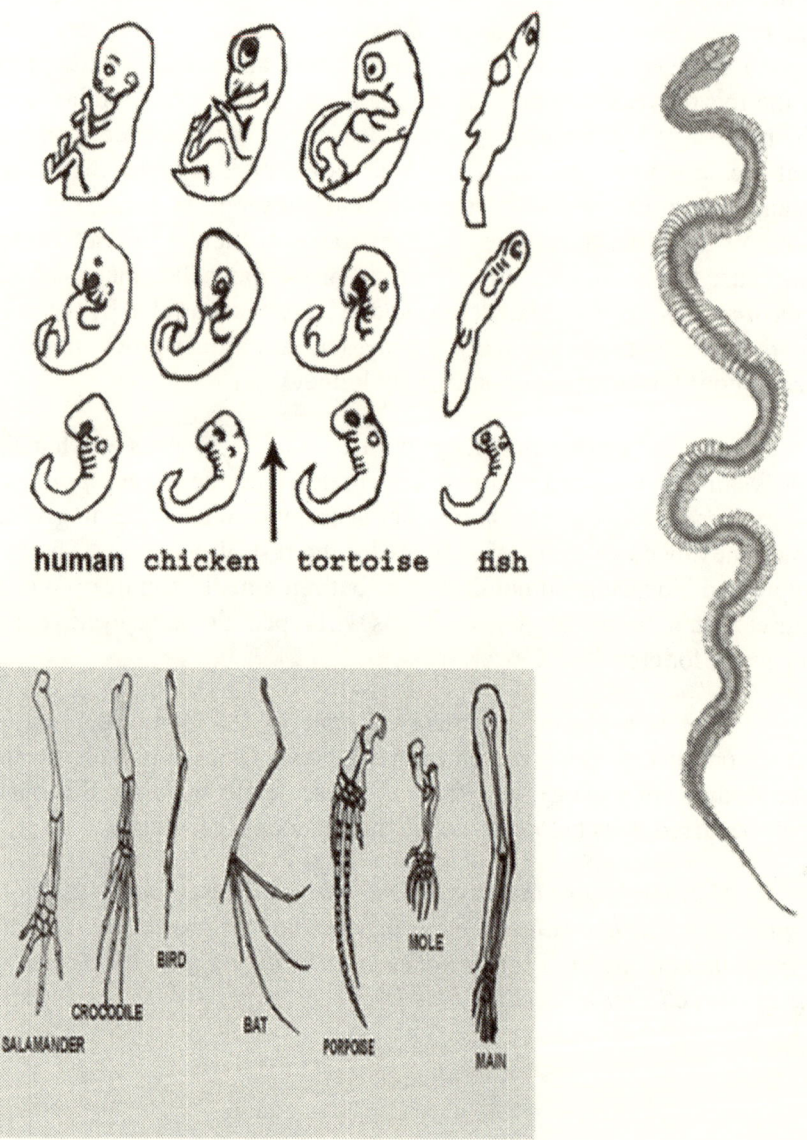

human   chicken   tortoise   fish

Figure (15), Human embryological evidence

There is an overwhelming similarity between the embryos of various members of the animal kingdom. But it lasts only for a few hours or days. It is as if they are trains which start from a single station, but soon each one changes course for a different destination. From an original blueprint of the forearm bones, various genes keep on lengthening some bones and shortening others, in order to create all these examples of 'homology', which means ancestral similarity. It is the same idea like the different shapes of dog-faces which we created by artificial breeding: Longer or shorter snouts, bigger or smaller ears, etc. Each specialized gene gets switched on and off at different times, and for different lengths of time, giving us the end results of different shapes of dog-faces.

Snakes are vertebrate animals, like humans. A certain gene, *common to both*, specializes in making ribs. In snakes, this gene is switched on for a much longer time during its embryological growth. The end result is that almost 90 percent of the snake's body is encircled by ribs.

Another well known and similar embryological reminder of our past is the occasional presence of accessory nipples, in males or females. They are always present along the so-called 'mammary line', which extends vertically from the middle of the clavicle. Humans usually have only one breast on each side, but most other mammals possess a whole line of them on each side, along that mammary line. Such congenital anomalies, which look like what takes place in lower animals, are collectively called 'atavistic congenital anomalies'. *They are simple reminders of our past, and of our suppressed genes.* In February 2000, the newspaper *"Sunday Times"* reported that American geneticists are discovering a gene in humans similar to the genes that induce some animals, such as arctic bears, to hibernate during winter for several months. The discovery of this gene could have intriguing practical applications. Funded by the Pentagon, researchers are investigating the possibility of *hibernating* severely injured fighters in war fields until they reach base hospitals. *SCIENTIFIC AMERICAN* Magazine [June 2005, pages 24 – 31] runs a whole article about this form of "suspended animation". All of this is a remarkable third millennium reminder of our genetic past. The more we research our junk or silent DNA the more we will discover the truth about the G - SAT theory.

CONTROVERSIAL QUESTIONS ANSWERED BY THE G - SAT THEORY

Does The Gene-Suppression-Activation Theory (G-SAT) fit with the scientifically proven facts of paleontology, Natural Selection, and geographical studies of the earth? Yes, it does. Not only that, it also sheds light on many points in Darwin's theory that have been unexplainable, and debatable, until today.

*** The G-SAT removes the need for twisted speculations to explain the weakest spots of evolutionist reasoning, detailed in Chapter 6.

*** The G-SAT beautifully explains the big question-marks in evolutionary history, such as Cambrian-age explosion of complex life, Punctuated Equilibria, Quantum Evolution, etc. By these terms, paleontologists mean the periodic pattern of abrupt and rapid evolution, followed by millions of years of relative stasis.

With the exception of Niles Eldredge, Curator of Invertebrates at the American Museum of Natural History, and Harvard Biologist Stephen J. Gould, the idea of punctuated equilibria is adamantly denied by most evolutionists. All evolutionists other than Eldredge and Gould say that further research will fill the supposed gaps of millions of years of stasis. In their view, the idea of punctuated-Equilibria could imply the *occasional super-natural interference, from time to time, by the Creator in the process of life evolution.* The argument, and its denial, has often been the subject of bickering in every conference or book dealing with evolution science.

*** The G-SAT beautifully explains the lack, in the fossil record, of what we call 'intermediate forms', or the missing links, between various species. For example, links between worms and crabs, between insects and fishes, between dinosaurs and birds, or between apes and Mankind. The question of these missing links is one big mystery in Darwin's theory. Argument regarding it has filled thousands of pages, in both evolutionist and creationist literature.

*** The G-SAT also gives a scientific recipe for reconciliation between the Ultra-Darwinists who claim the supremacy of what is good for the genes, and the Naturalists who focus on whole creatures and species, rather than genes.

*** The G-SAT beautifully explains the paradox of domestic animals. Dogs, horses and camels have the *genetic* ability to become domesticated and trained. Their *trainability* is a strong instinctive character, but this character is supposedly connected to the presence of the *training species*, which is Mankind. These animals *preceded* Mankind by some 100,000,000 years. The survival value of this trend as an isolated evolutionary process does not make sense. *It had to be a part of a pre-determined master-plan.*

*** The G-SAT beautifully explains the astonishing similarity between mammalians evolving separately on isolated continents. Anteaters, moles, flying squirrels, cats, and wolves, as well as their marsupial versions, all have evolved in complete isolation on continents that have never been joined. And in spite of that, there are lots of amazing *look-alikes* everywhere. Parallel developments of many similar species have taken place in the Old World, the New World, and Australia, totally isolated from each other. The genes for each of these varieties were always there, ready for activation whenever the environment is ready, giving specialized signals in every case.

*** The G-SAT beautifully explains the sets of genes which act together in a complimentary way. Examples are the sets of genes which create an herbivore or a carnivore, giving it special teeth, stomach, guts, claws, sense organs, etc. Another example is the team of genes which create the highly sophisticated ability of echolocation, including sonar emission, sonar reception, special nose, special ears, brain programs, etc. And last but not least, the team of genes involved in the human acquisition of language, including larynx, tongue, lips, nose, and the language syntax in the brain, etc.

*** The G SAT beautifully fits with the all-or-none-rule

The application of the all-or-none concept in general extends far beyond the hemoglobin and chlorophyll molecules as explained in Chapter 6. The balances of the various organs of a complex body such as the mammalian body, or any living animal, insect, or plant, is *all-or-none*. You can't have a viable body with only an efficient circulatory system, but incomplete or deficient digestive or nervous, or renal or respiratory system for example. All the organs of a complex living creature must be working in a complimentary and harmoniously balanced way from *day one*, if it is to grow and keep alive. This is an all-or-none prerequisite.

Now let us contemplate this obvious fact with the sudden appearance of complex life forms in the Cambrian Explosion of life 540 million years ago. It becomes evident that when each life form did start its existence, it had all its organs well balanced in function already. There can't be trial-and-error improvements of the balanced function of the various organs of a complex living system. Any imbalance is incompatible with life.

From that moment on, in competition with its prey or predators, further improvements can take place, for example muscle strength, running speed, shape of teeth among others. This is natural selection, or micro-evolution. This is totally different from macro-evolution which is the original emergence of the life form, in other words: species, in its basic anatomy and physiology. If it succeeds in its Natural Selection endeavors this life form, or species, will multiply and thrive, if not then it will become extinct.

*** The G-SAT beautifully gives a feasible explanation for another sticky dilemma. Evolutionists have difficulty trying to invent a survival value for some of the weird habits of Mankind. What is the survival value of our space programs and our relentless study of the Galaxies and the Universe? What survival value is there in our appreciation of music, beauty and arts? What survival value is there in the basic human instinct of private ownership?

*** The G-SAT removes the factor of randomness, haphazard chance, and blind watch making, which has been the weakest link in all evolutionary speculations. Directed-Natural-Selection would be comparable to Guided-Artificial-Selection. The latter has produced its results in the short life time of Mankind. The former has produced its effects in the relatively too short time of the history of life on Earth, *(too short for haphazard-chance).*

*** *The G-SAT gives a scientific recipe for reconciliation between evolutionary facts, and creationists' beliefs. How is that? The Creator has set this clock-like mechanism, and put it on, 3,500 million years ago. It keeps on ticking through today, with its products getting fine-tuned by Natural Selection all the way.*

Simply, this is the principle of Deism. Deism was originally a philosophical idea initiated in the 16th century. Same as any philosophical thought, it all depends on the eloquence of its champions, with no science involved. The G-SAT theory gives a scientifically feasible flavor to the whole idea. For the first time in human history, both Science and Faith could have a single starting point.

THE DYNASTY OF ATOMS

Genes are protein molecules, molecules are made of atoms, and atoms are made of subatomic particles. Modern science has proved that subatomic particles were the *only* form of matter generated at the fateful moment of the Big Bang. Within less than a minute, of our time, the Universe became a huge ball of fire. It was millions of light-years in diameter, and contained nothing but the atoms of hydrogen and the atoms of helium. From that minute, onwards, those atoms look to be strictly observing certain physical rules, which are valid till today: electromagnetism, gravity, and strong and weak nuclear forces. Under the guidance of these physical laws, those simple atoms have gathered to form more complex ones.

With the passage of billions of years, and still under the strict guidance of these physical laws, they have formed a planet called Earth. One and half billion years later, *doubtlessly still under the guidance of these physical laws,* they have formed complex molecules called DNA. DNA has on it certain coded instructions called genes. DNA and Genes are the legitimate descendants of 'The Dynasty of Atoms'. In that dynasty certain physical rules have been strictly observed all the way. *These Laws of Physics did not evolve nor change, were they just given? If given, then by whom?*

This is a simplified presentation of a hot and sophisticated scientific argument termed *'Biological Determinism'.* Its proponents reluctantly admit that it is tantamount to a miracle in nature's clothing. It simply means that the course of shaping of the universe followed by initiation of life were *both* divinely ordained. Representative books on this thesis are [de Duve, 1995], and [Davies, 1999]. In his book, page 147, Physicist Paul Davies gives a very interesting calculation by saying that contemporary DNA is the same DNA that almost certainly existed 3500 million years ago. Each DNA molecule contains billions of atoms. When an organism dies and decays, its atoms are released back into the environment. Some of them eventually become part of other organisms. Simple statistics reveal that my body contains about one atom of carbon from every milligram of dead organic material more than a thousand years old. This simply means that I carry a billion or so atoms that once belonged to Tutankhamen, or Aristotle, or a tree in the gardens of ancient Rome. Looking at my body this way, I should reflect on the long and eventful history of its atoms, and remember that the flesh I see, and the eyes I see them with, are literally made of atoms that were cooked billions of years ago in the kitchen-work of the stars.

Another name for all this is the Principle of Deism: God has set the clock of life, as part of the physical laws validated at the beginning of the universe, and then just let it go.

* * * * * * * * *

POSTSCRIPTUM TO CHAPTER (7)

I very much enjoyed, scrutinizing a book by the Nobel laureate and 'co-father' of the DNA discovery, James D. Watson. The book is titled *"DNA: The Secret of Life"*, [Watson, 2003].

In his book, on p.35, Watson gives credit to a book titled *"What is Life?"* written in 1944 by a physicist called Erwin Schrödinger. Its author believed that life is a hereditary code-script embedded in the fabric of the chromosomes of every living cell. The mere idea was the spark that hooked young James D.Watson to the road that led him in 1953 to the discovery of DNA's Double-Helix, and Nobel Prize.

Watson uses another name for junk-DNA as 'Introns', long segments of the DNA chain which do not code for the manufacture of any protein. One theory holds that these molecular intruders are merely vestigial, an evolutionary heirloom, left over from the early days of life on Earth. Still, it remains a much debated issue how introns came to be, and what if any use they may have in life's great code.

On p. 131 he asserts that natural selection takes eons, but directed molecular evolution can do a job in just hours or days.

Like the G-SAT theory, Watson further states, (p. 264), that we now know that many of the most interesting parts of the genome lie outside the genes, constituting the control mechanisms that switch the genes on and off. Additionally, on p. 265, he comments that the chromosomes of humans and chimpanzees are very similar. Chimpanzees, however, have 24 pairs whereas we have only 23. It turns out that our chromosome No. 2 was produced by the fusion of two chimpanzee chromosomes. In 2002, a gene named 'FOXP2' was found to be closely related to the genetic language syntax. Mutations in the human version of that gene have been found to cause linguistic impairment. Out of a chain of 715 amino acids in this gene, just two changes distinguish humans from chimpanzees and gorillas. Actually the amino acids in the gene 'FOXP2' are identical in chimpanzees and gorilla, as well as all other mammals tested, except in humans.

I was also very pleased to come across a recent issue of *"SCIENTIFIC AMERICAN"*, Oct. 2004, in which there is an article titled: 'The hidden genetic program of complex organisms'. The subtitle reads as such: "RNA 'junk' inside cells may directly regulate how a fertilized egg becomes the trillions of cells in a human body."

One paragraph of the article is summarized as such:

" ….. This revolutionary view of complex life sees the DNA genes as islands of protein manufacturers in a sea of regulator RNA. Indeed, these results suggest a general rule with relevance beyond biology: organized complexity is a function of regulatory information. The explosion in complexity of life that took place half a billion years ago took place as a result of advanced controls and embedded networking of the genetic code…. "

On page 37 it says: "… What was missed as 'junk' because it was not understood may well turn out to hold the secrets to human complexity and a guide to the programming of complex (life-) systems in general …."

The article confirms my prediction that the scientific validation of The G-SAT theory presented in this Chapter 7 could be around the corner in less than a decade.

* * * * * * * * *

CHAPTER 8

* * * * * * * * *

NOAH'S FLOOD

* * * * * * * * *

In the history of Mankind, the story of the flood became an outstanding landmark between what was before, and what came after. The story is present in almost all cultures, in some way.

One and half centuries ago, Darwin suggested an ancestral relationship of all forms of life. Still, one obstacle had to be removed. That obstacle is the *literal* way Religious Scriptures were being interpreted by the clergy in *all* Religions, Monotheistic, Hinduism, Shintoism, Buddhism, etc.

The main points of difference are three, the time scale of Genesis; the original creation of every animal in its present form; and the story of the Great Flood; during which someone used a single boat to save *all* the forms of life on the whole Earth. Let us discuss each separately.

TIME SCALE OF GENESIS

The symbolic interpretation of all these points is the only explanation that would be scientifically feasible. Nobody would agree with Reverend James Usher, the Irish Bishop, who declared that his calculations, based on Biblical Stories, show that the Creation Day was the 23rd of October in the year 4004 BCE. Neither would anybody agree with the British naturalist named Phillip Henry Gosse. In 1857 he thought he had solved the conflict between the clergy's belief, and the scientific fossil findings. He suggested that God created Earth and all species in their present form in the year 4000 BCE, and has also simultaneously created these fossils and made them look like millions-of-years-old, just to test our faith. He said that all our paleontological and geological findings are mere appearances of pre-existence, fashioned by God at the same moment of creation, 6,000 years ago.

The clergy insist on the seventh day of Holy week, Sunday, Sabbath, or Friday, and the literal inference they give is that the six days of creation were days of our time scale. We now know that time is purely relative. The faster an object moves in space, the slower will time proceed for it. If the object reaches the speed of light then time will stop for it altogether. Surely a *day* in creation scale could be millions of years in our Earthly perception.

CREATION FORMS

The present forms of animal life can never be the same forms which existed millions of years back, they have gradually changed. Paleontological studies are adamantly clear about this scientific fact. Millions of years of changes of habitat, weather, and natural disasters have enforced their fingerprints on all forms of life, by the mechanisms of natural selection. Millions of species have simply become extinct. Other millions have greatly changed, in order to accommodate to the changes in their environment, predators, or prey. There is not a single verse in any Religious scripture that specifically states otherwise. Thus the clergy's insistence on such inference is totally unjustified.

NOAH'S FLOOD

To Noah, the great flood must have been a hideous drowning of *his* world. This could have been a whole valley with several towns. There is not a single verse in any Religious scripture that specifies the *whole* world. Noah's ship, however large, could have never accommodated even a fraction of the millions of species of animal life on Earth. It could have accommodated the animals of their *limited world.* Only the symbolic rather than the literal interpretation of the story of the flood could make sense. It would symbolize the omnipotent power of the Creator to save the believers from Nature's wrath. Since the emergence of Mankind, which is very recent, there have been geological and paleontological signs of many local floods, but none which drowned the *whole* Earth.

In a previous ice age, the Atlantic Ocean level dropped well below the Gibraltar strait level. The Mediterranean water supply was cut off, there was no Suez Canal, and then it gradually dried up because of evaporation in one thousand years. It is calculated to evaporate at the rate of 40,000 cubic kilometers of water every year. With the end of that ice age, five million years ago, the Atlantic Ocean began to rise and to pour down again at Gibraltar at a rate 10,000 times faster than the present flow rate of Niagara Falls. It filled the Mediterranean back to its original level in less than one hundred years. The torrential water eroded the Gibraltar connection to its present width, and eroded it down to more than 1,000 feet. This depth is beyond any later drops in sea levels by later ice-ages. So this Mediterranean incident was never repeated again.

A book published in 1999 titled *"NOAH'S FLOOD"* was written by two oceanographers from Columbia University, William Ryan and Walter Pitman. The authors used its 320 pages for an extensive study of the legend as well as the scientific facts. They believe that 7,500 years ago the rising water in the Mediterranean broke through a natural dam at the Bosporus Strait, rapidly expanding the Black Sea by more than half. It drowned a flourishing civilization on its shores, and forced them to migrate both northward into Europe, and southward into the Middle East.

The research techniques they used were the most sophisticated 20th century scientific tools in the fields of oceanography and hydrography. They were funded and manned by scientists from prestigious European and American universities. They included studies of ocean levels across millions of years, studies of the faunas and sediment in the ocean and sea floors as well as drowned river-beds, studies of earth's magnetic field reversals and their manifestations on land and under water. It also included sophisticated satellite photographs and scientific measurements.

These studies reached the definite conclusion that 7,500 years ago the Black Sea was a fresh water lake. Its fresh water was supplied mainly by four rivers: the Danube, the Dniester, the Dnieper, and the Don. This Black Sea Lake was almost half its present size; but more importantly its level was 400 feet *below* the Mediterranean Sea level.

In its time, the Black Sea was the largest fresh water pool in the region. The surrounding areas in Europe and the Middle East were, at that time, dry barren and arid stretches of land. The population around this fertile fresh water source created an advanced civilization. They mastered the arts of agriculture, fishing, and animal domestication. They stopped being nomadic foragers, and became farmers and herders. They built houses and villages, and were on their way to an industrial civilization.

Suddenly, 7,500 years ago, the natural barrier at Bosporus broke down. One factor was the rising Ocean and Mediterranean levels at the end of the last ice-age, which was already 400 feet higher. Another precipitating factor was the frequent earthquakes in the region. A torrential flood of sea water poured over the brim, at the rate of 200 Niagara Falls. It eroded the new connection between the Mediterranean and the Black Sea to its present width, and to a depth of around 200 feet.

The foaming mist, the possibly simultaneous heavy rains, and the roaring sound of that flood must have terrorized the whole valley for hundreds of miles. People near the Bosporus must have witnessed a most frightening scene, and soon spread the word everywhere. The people living around the lake saw the water edge advancing almost a mile every week, drowning their shoreline and their villages. The water level was *visibly rising by two feet every single day.* The calamitous inundation continued unabated for 300 days, until the Black sea became level with the Mediterranean. Its size has already doubled, and has drowned the Azov valley, south of Russia, creating the present-day Azov Sea.

In less than a few months, a whole civilization was drowned. The main harm was not the physical drowning per se, for which there were enough warning signs to force people to pack, carry their valuables and cattle and get away. The more painful harm was the death of all fresh water fish, the death of all the grass fields needed for their animals, and the disappearance of their farmlands. Their fresh water supply was replaced by salt water which has even drowned the above mentioned four river-bed deltas and climbed up-river for several miles. That salt water was useless to humans, cattle, and most importantly to their seeds. The only explanation they could think of was that this is the action of angry God or Gods, in punishment for their sins. Mankind of those millennia could never comprehend the mechanisms of Nature's catastrophes.

Within a few weeks from the start of the flood, the inhabitants of the area realized the magnitude of their loss. *They will have to leave home for good, in a massive exodus into the unknown.* As farmers and herders they instinctively saved the only possessions that could sustain them in a new land. These were their domesticated animals and birds, their seeds, and their tools. It was these very things that their *Sages* stocked in the big ship, Ark, or ships, that they rapidly built.

This was their best possible means of escape, rather than running on foot for an unspecified distance or time. They would build a ship, load it with all the necessities for future sustenance, and then wait for the rising water to carry them for as long as the water rose and flowed, and as far as the water will ultimately reach.

For those refugee populations, the flood event became a historical landmark that divided their world into what was before, and what came after.

According to one interpretation of the Old Testament, Mount Ararat in eastern Turkey is mentioned as the site where the Ark of Noah came to rest after the Flood. It is interesting that it is the highest mountain in the area, surrounded by several small lakes, and it is only within 200 miles from the shores of the Black Sea. Mount Ararat is 17,000 feet high and its peak is permanently ice-capped. Another story claims that the site was a smaller mountain, (7,000 feet), southwest of Ararat.

For the Flood to get that high means an in-between-the-lines message that it drowned the *whole* earth, up to 17,000 or 7,000 feet above sea-level. *There is not enough water or ice on planet Earth to drown it that high*. Also such a universal inundation has never been detected or proved. Some clergy argue that the finding of marine-life fossils high up in mountains may be a proof of just that inundation. But this is not true, because these same mountains are proved to have been raised from the sea bottom, millions of years ago. It was caused by the movements of the earth's tectonic plates and drifts of the various continents explained in Chapter (3).

The colossal migration of the refugee tribes that inhabited the Black Sea shores carried them to Mesopotamia, the Near East, and to central Europe. Linguistic studies have shown a remarkable linkage of common words, and phonetic and grammatical constructions among many languages. These languages include: Sanskrit, Latin, Greek, Welsh, Early German, Semitic, and Persian. They could not possibly have been produced by accident. The overlap strongly suggests that these separate languages have sprung from some common source.

The legend of the flood in the Rig-Veda Hindu texts is similar to that in the Gilgamesh Epic of the Babylonians, and both are identical to the Flood story in the Old Testament. The story details in all these legends leave no doubt that *all are describing the same historical event*. The Holy Sage who understood the warning signs and started to build the Ark was given different names in the various legends: Manu in the Rig-Veda; Atrahasi or Shuruppak in the Acadian stories; Ziusudra in the Sumerian; Utnapishtim in the Gilgamesh Epic and; Noah in the Old Testament. Before the invention of writing, history was passed from one generation to another by word-of-mouth, in songs, hymns, epics, and legends that crossed the centuries almost unchanged.

The scientific data that came out of this research around the Black Sea, and under its waters, are proven beyond any doubt. Their implications have a perfect fit with a human drama that took place in the year 5,500 B.C.E. that is 7,500 years ago.

In the year 2002, another book *"Before the Flood"*, [Wilson, I. 2002] describes further research, as well as Titanic-like diving robots, that have confirmed the remains of towns, villages and stone structures of an old civilization under the stagnant waters of the Black sea.

> > > > >

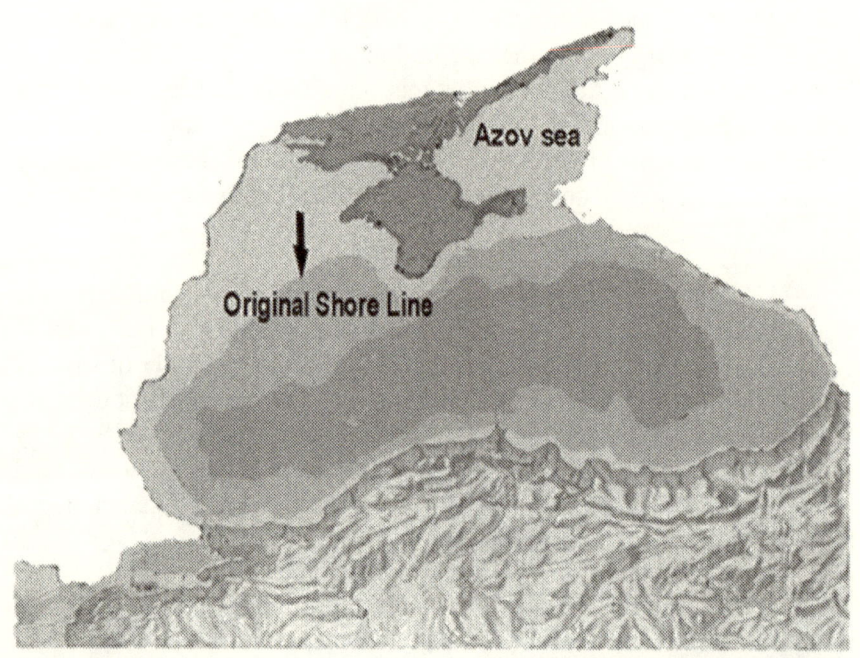

Figure (16), The Black Sea, the drowned cradle of civilization

CHAPTER 9

* * * * * * * * *

MANKIND

* * * * * * * * *

"Look at man's pink skin and puny frame -- how could natural selection evolve greater weakness without the counter-balancing gift of reason being bestowed first? Man must have had the human proportions of mind before he could afford to lose the bestial proportions of body."

Charles Darwin.

"It would appear that around 50,000 years ago, modern humans suddenly became culturally modern: we see in the remains from this time the first indisputable ornaments, the first use of bone, ivory and shell to produce familiar and useful artifacts, and the first of many improvements in hunting and gathering technology. What happened? We shall probably never know."

James D. Watson.

At the end of this chapter, I'll tell you a story about aliens who visited Earth as Gods, and implanted wisdom and intelligence into Mankind about 30,000 years ago. It is remarkable that such falsified stories are believed by some human audience. It defies explanation.

Hussein A. Amin.

WHO IS MANKIND?

There are certain facts about you and me that nobody ever disputes. First, Mankind is a one and single species. Love is *fertile* between any girl and any boy, yellow, white, dark or black. Second, Mankind *is* different from all other forms of life. The differences were defined in many ways throughout history, sometimes in slogans, sometimes in funny words. Some examples are: The Naked Ape, the Talking Animal, the Smiling Animal, the Intelligent Monkey, the Conscious Animal, the Beast with a conscience, the Artist Animal, the Altruistic creature, the Tool-using animal, etc. Philosophers and scientists have endlessly squabbled over the definitions of these differences, as well as their causes. Further, some scientists have even challenged the monopoly of Man over many of these characteristics. Some of them have tried, and failed, to simulate human language to the voices, songs, and warning shrills of apes, birds and animals. Others have spent their whole careers vainly trying to find an even faintly comparable degree of intelligence in the animal world.

THE ONLY TWO EXCLUSIVE DEFINITIONS OF MANKIND

First, Man is the only animal who can invent new ideas from previous information. This momentous ability has enabled him to change the world around him to suit his frail body, rather than accommodate his body to suit the environment, or the world.

Second, Man is the only animal who, by using language, can transfer to his descendants any new experience or knowledge he gains. A human descendant does not start from square one, and end at the same square where his father ended, as all animals do. He builds on his ancestors' experience further and further, in an endless progressive gain.

Let me start by examining Language. Language use *IS* monopolized by Mankind; there is no doubt about that.

The term language is not the dictionary you pick up in a library. A dictionary is only a list of the vocabulary of any particular language. There are hundreds of dictionaries for hundreds of living and dead languages, as well as hundreds of vocabularies for Creole languages which have no dictionaries at all.

By the term language it is meant: the way we express a sentence which describes the incident of *"who did what to whom, and why, when, and how?"* The verb did and the three nouns, who, what, and whom, are the four basic elements of what scientists call *'The Syntax'* of language. In the last ten years it has been proved beyond doubt that language syntax is a genetic characteristic of the human mind, it is one of the basic instincts of Man.

The language instinct in Mankind has no direct relationship to the larger size of their brain. Brain size per-se is not a factor in the ability to learn language. Patients with microcephaly, congenitally smaller heads, may have brain size of only 600 cubic centimeters, almost the same size of the brain of a gorilla. They do have very limited intelligence, yet they can use and learn language in a human way.

The genes for language lie dormant in the newborn infant. They get switched on at the age of two years. Children get this remarkable ability between 2 to 3 years of age, without having to be taught how to express their thoughts. The *when* is the last of the abilities to be achieved, usually at the age of three years. It refers the timing of the incident to the present, future, or past.

What the children acquire from their family is only the vocabulary. A French girl will call the dog "Chien", an Italian will call it "Cane", and an Arabic girl will call it "Kalb". Even in sign language, although a different shape of vocabulary is used, yet the syntax rules are still the same.

The human child is born with his brain already wired, ready for language. The brain centers of language express themselves through a *genetically inherited* very special arrangement of the nerves and muscles of the tongue, larynx, lips, and nose. The whole set-up is an inter-complimentary, *instinctive* system, present only in our species.

ARCHAIC PAINTINGS AND CARVINGS

Spoken words disappear into thin air, but written words do not. The earliest man recorded his thoughts for his descendants in form of pictures and paintings. Examples are the famous paintings of the caves at Lascaux in southern France, and in many other parts of Europe, (see Fig.

17). The ages of these paintings have been specified, some are fifteen, and others are twenty to thirty thousand years old. They show their painters as really sophisticated humans, much more sophisticated than the layman idea of savage cave-men. The painters were forced to live in caves, which protected them from the harshness of the then prevailing ice age. They used fire to warm the interior of the caves, and to give them enough light to paint these beautiful pictures. Recent studies suggest that these pictures were mainly lessons from the elderly to the younger generations, in the art of hunting animals for food. The art of carving is also another form of expression, which needs unquestionable inventiveness. One of the oldest discovered samples is a sculpture of a female figure in the rock shelter of Laussel, Dordogne, France. Carved from a limestone slab aged 27,000 years, it shows a fertile woman with fatty hips and hanging breasts. In her right hand she holds a crescent-shaped object on which 13 clear lines were itched. The more you contemplate this sculpture, or its pictures, the more you believe it was a demonstration by the cave medicine man to his students. He taught them about the menstrual cycle, and its relationship to the 13 lunar months of each year. [James and Thorpe, 1994, page 486.]

Another discovery was a collection of 17 ivory statuettes found in caves in Swabian Jura, south West Germany. They dated 30,000 years old. They depicted animals such as mammoth, bison, bears, horse, cats, and even a composite statue showing a man with a feline head. This is a comment from the book *"Dawn of Art"*:

> "These statuettes show that the human artists of that age knew very well how to work ivory, and make beautiful figures from it. They also had the full mastery of graphic technique". [Chauvet, 1996, page 126.]

May I also repeat that these people were far from savage; on the contrary, they were as inventive as we are today. The only difference between us and them is that today we stand on top of a great mountain of scientific knowledge, which allows us a much greater field of vision. We didn't inherit this mountain out of nothing; it has been built up, a bit by bit, starting from 30,000 years ago.

CREOLE LANGUAGES

The languages of our cave ancestors were surely the same as the Creole languages of today. These languages are used in faraway islands and isolated communities. Using their genetic syntax of language the people invent a name for every object they see. Their invented vocabulary becomes a very special language that nobody else understands. Many of these Creole languages have no written symbols, they are transmitted from generation to generation only by word of mouth, as well as the hand and head signs during the talk. In our study of the history of contemporary major languages we find that they all have started this way.

WRITTEN LANGUAGE

The next essential step was to specify a certain inscription or mark for every sound or word. From that idea writing was gradually developed. This gave early Man a better way of transferring his gained knowledge and experiences to his descendants. The earliest discovered record of written words dates to about 10,000 to 12,000 years ago. They were surely the first ever written words, because they were extremely primitive, in the form of scratches, lines and angled marks. The Hieroglyphics of the ancient Egyptians came six thousand years later and were much more sophisticated. It was a 'pictographic' writing, using drawings of animals, houses, people, etc., in place of letters.

Figure (17): Archaic cave paintings, The painters were as sophisticated people as we are.

The first written languages were in form of pictorial letters, each letter signified a whole word, or even a whole sentence. Later came the phonetic letters, each letter signified a specific sound. Most modern languages have phonetic letters.

From this we can confidently specify the age of written language at 12,000 years, and the age of painted or sculpted expression at 30,000 years. **The importance of these remarks is that they also specify the time Mankind emerged as a very unique type of species, separate, and different, from all the rest.** Man became capable of transferring experience, knowledge and habits in verbal and recorded form. Thus he was liberated from the tyranny of the 'genes' as the sole connection between the future, the present and the past.

INVENTIVENESS

The foremost character of Mankind is their *inventiveness*. I prefer this word rather than the mind, or intelligence, of Mankind. The last two words have been extensively degraded from overuse. Everybody talks about the 'mind' of a dog, and that of a gorilla. Everybody compares the 'intelligence' of a chimpanzee to that of Man, and to that of artificially intelligent computers, abbreviated as AI, for Artificial Intelligence.

Curiosity, inquisitiveness, creativity, the love of adventure and discovery, taking risks, among others are all characteristics that sharply dissociate us from other species. But *inventiveness* is far more descriptive, and far more specific, to the unique species of Mankind. It is the ability to deduce new and complex ideas from previously known simpler facts. Using this magic-like ability, Man has reached into limits far beyond the wildest dreams of his ancestors. He has changed the World around him to suit his frail body, rather than migrate or evolve as birds and animals do, in order to suit their habitat. All the inventions of the industrial revolution were about augmenting human physical and visual abilities. Then, 50 years ago, came the computer. Computers enable us to try to extend human mental capacities as well. Most important of all, Mankind has started to ask questions about *how* he, and the entire Universe, came to be, and *why*?

Mankind has done all that in a mere 300 centuries, a mere blink of an eye in the history of the Earth. At the same time all other species of animals have timidly continued to live exactly as their ancestors used to do for hundreds of millions of years. Even when some animals learn certain intelligent tricks, such as circus animals or police dogs, we find them neither willing, *nor able,* to transfer this experience into their offspring.

BRAIN SIZE, A FALSE MEASURE

As mentioned above, the small brain size of patients with microcephaly does not affect their acquisition of language in a normal human way. The pioneers of paleontology concentrated on 'brain size' as a supposedly reliable measure of intelligence in the animal kingdom, including Mankind. They made endless comparisons of the brain size of Mankind with that of various fossilized apes, claiming that those apes were *'early humans'*. Paleontologists even gave those apes names such as 'Homo' in order to emphasize their ancestral relationship to Mankind. In my view, the origin of this bad habit is the insistent tendency of many paleontologists to avoid admitting the *'sudden'* emergence of Mankind 30,000 years ago. Instead, they claim a much longer history of millions of years. The definition of Mankind is the inventive and talking animals, this did not take place before 30,000 years.

Paleontologists claim that our intelligence and our language have originated by the concerted actions needed for hunting for food. I think this claim is completely unjustified. You need only see, and enjoy, one of the magnificent video films of David Attenborough to see that such claims are utterly false. In a film titled *'Trials of Life'*, you see the concerted action of a dozen male chimpanzees, driving their victim into a certain spot on tree tops, and then catching it. The victim is a Colobus Monkey, which is then brought to the ground where it is torn to pieces. The excited family of females and youngsters are then invited to the meal. It is also common knowledge that lions and wolves often flush their prey toward a concealed member of the group. The hunter chimpanzees, lions, and wolves have been doing this for millions of years, without having to talk, and without becoming as intelligent as we are.

FOSSILS OF BURIED INTELLIGENCE

The old signs of inventiveness can easily be looked for, but first they have to be specified. One of the most important characteristics of Mankind is our vulnerable bodies. Our naked body can easily be frozen to death; it must be covered by warm clothing, made of animal skin and fur. Our legs are not fast enough to catch prey, or escape predators. We have to augment our speed and strength by the use of missile weapons such as spears, arrows and boomerangs, and later guns. We can't endure hot suns or heavy rains, we need shelters and houses.

So when we search for the first signs of inventiveness, *as the first sign of Mankind*, we should be specifically looking for the fossilized remains of these three particular items, namely clothes, weapons and houses. These have been unearthed in many parts of the world, and their ages do beautifully fit with the limit of 30,000 already specified by the paintings and art objects.

In the book "*Ancient Inventions*", [James and Thorpe, 1994], the authors on page 380 describe the oldest boomerangs in the world, dating 21,000 years. They were found in Oblazoa Rock Caves in southern Poland. Made of wood, each has a span of 28 inches. The earliest flint arrowheads dated at 18,000 years, have been found at the site of Parpallo in eastern Spain. The oldest known fishing hooks date about 14,000 years, they came from Stone Age sites in Europe and South Africa.

As to the clothes, we now know that the hunters who colonized the frozen north during the Old Stone Age did not dress in furs casually slung on their shoulders. They needed well tailored garments, sewn from animal skin, and tightly fitting around the body. Archaeological sites dating as far back as 22,000 years have produced eye-needles made from animal bones. Their purpose was surely to make clothes. In Sunghir, Russia, in 1964, a burial site of three men dating 20,000 years was discovered. Each of the men had hundreds of ivory beads on his head and ribs. They were the remnants of bead decorations on their shirts and caps. At a contemporary site at Buret, in southern Siberia, a figurine carved from a mammoth ivory tusk was found, and dated at 20,000 years. It is 4.8 inches high, showing an individual wearing a fur suit very like the traditional Eskimo costumes. The suit covers the head as well, leaving the face in front, with eyes, nose and mouth beautifully carved.

The last ice age ended only 10,000 years ago. This explains why we find most of the human artifacts before that age inside protective caves. With the recession of ice, the early inventive humans could venture into the open. Only then did we start finding the archaeological evidence of shelters, houses and cities. These ancient shelters today form the rich wealth of tourist attractions, in Mesopotamia, in Greece, in Egypt, in Europe and in South America.

DEFINITION OF MANKIND

Mankind is defined by their only two exclusive characteristics, language and inventiveness. The oldest evidence of these two was suddenly started 30,000 years ago. We can confidently say that Mankind is a species which *suddenly* appeared on the surface of the Earth 30,000 years ago.

In paleontology, there are certain fossils which we call *'The Hominid Line'*. They were found mainly in Africa, and were given various names according to the discoverer or the location. The most famous one was called "Lucy", an almost complete skeleton of a female, dated at 3 million years old. The two main characteristics given for the hominid line are *bipedalism*, which means walking upright on two feet, and changes in the four canine teeth, being smaller with sharp incisor-like edges. The argument is that these are the earliest ancestors that evolved into Mankind. In spite of that argument, they surely do not belong to us; they represented other species of apes. They were not naked; their muscles were much stronger and massive. Their tool usage is comparable to that of otters, monkeys, and Egyptian-grey-hawks, (see page 182).

In my view, this sudden landmark of 30,000 years is of highly important significance. In evolutionary terms it is almost instantaneous. The molecular clock of evolution, theorized by Motto Kimura in 1968 recognizes the average transition time from one species into another at about quarter of a million years. The way we appeared on the stage of Earthly life does not fit into the recognized evolutionary trends of the animal world. Animals feed and fight through instinctive behaviors, automatically run by specific genes. They do not actively think up ways of food, protection and procreation, they just obey their instincts. None of this is represented in Mankind. As Charles Darwin cleverly noticed, it is a most remarkable coincidence that, at the moment Mankind appeared on the Earth, they simultaneously owned the most vulnerable bodies in the animal kingdom. Their inventiveness plus language became the only compensation for their poor physique.

I can safely state that *"It looks as if some ingenious force has suddenly shaped Man's body that vulnerable way, endowed him with language and mind, and then charged him with the responsibility of feeding, protecting and fending for himself."* Looking back over 30,000 years it is amazing to see how Mankind did succeed in taking charge of this responsibility. Not only did we protect and feed ourselves by our own ingenuity, but also we conquered the rest of the animal kingdom, and then ruled the Earth.

The suddenness and the simultaneity of development of these exclusive characteristics, vulnerability, language and inventiveness, point to the creation of Mankind by a process totally different from the present concept of how Evolution works.

This incident took place during the most recent ice-age, which lasted 100,000 years and ended 10,000 years ago. It is as if at a certain moment, in the darkness and coldness of that ice-age, Homo sapiens got a few of its genes suppressed and few others activated and *VOILA!* Man became mysteriously transformed from a hairy muscular primate into a miraculously new creature. His body became naked and much weaker, and he became intelligent and talkative. In scientific language he acquired a second adjective: Homo sapiens, *sapiens.* In ordinary literature he was called Mankind. This metamorphosis is almost comparable to the worm which gets cocooned for a few days, and then what comes out of the cocoon is a different creature: a butterfly, with wings, long legs, and sex organs.

The mystery of this Sapiens transformation may never ever be solved. We have to choose one of three explanations. The first is the supernatural creation of a new species. The second explanation is the legend of aliens visiting the Earth, 30,000 years ago, and inseminating it with their offspring. That odd idea is given more detail later, under the title of *'Was God an astronaut?'* The third explanation is the scientifically plausible G - SAT theory, explained in Chapter 7. Guided natural selection has finally fulfilled its ultimate aim by creating us, to inherit, and to rule Planet Earth.

It is a pleasure to find many authors agreeing with me on this 30,000 years time-mark. Graham Hancock on pages 30-31 of his book *"Supernatural"* says that there was a great time gap between the appearance of the 'modern' skeleton of Mankind 190.000 years ago and his 'modern' behavior. At about 40.000 years ago Mankind underwent an immense behavioral metamorphosis, becoming innovative, artistic, symbolic, cultural, religious, and self-aware. That new brain function was not accompanied, or immediately preceded, by any obvious anatomical change. Hancock calls this as "the greatest riddle of archaeology". [Hancock, 2005]

Trevor Williams on page 14 of his book *"History of Invention"* writes:

> " .. To find men with whom we ourselves could identify we cannot go back more than say 25 thousand years, although Homo sapiens developed much earlier. A tool-using primate is not us,

civilization meant much more than that. It involved further two important capacities: (a) rational conceptual thought, and (b) language syntax. No signs of these two capacities are detected before 25,000 years ago." [Williams, 1987]

> > > > >

THE HUMAN MIND AND SOUL

You may have noticed that I have completely avoided the word Soul throughout this discussion. I believe that, Soul or no Soul, Mankind is defined by their only two exclusive characteristics, language and inventiveness. These two characters are only 30,000 years old. This suddenness is completely different from the way Natural Selection and Evolution work. It does not squarely fit into Evolutionary Science, and cannot easily fit into the gradualism of Natural Selection.

Let us suppose that scientists could ultimately prove that Rene' Descartes was wrong. After all he was a mere philosopher, not a scientist. Let us suppose that they could ultimately succeed in proving that our consciousness, memory, subjective feelings and intelligence are actually affected by purely physical and chemical reactions in our 100 billion neurons; SO WHAT? This is a pure technicality, which explains nothing more than that. Could you claim that the silicon chips in your computer are the source of all that wisdom visible on your screen?

The genes for language and inventiveness were suddenly activated 30,000 years ago. The genes for language and inventiveness are THE definition of Mankind, they ARE Mankind! Mankind is not defined as the animal that has a Soul, but as the animal that can invent ideas, and then talk about them, (see figure 18).

According to this theory, the sudden, and simultaneous, activation of two forces of that magnitude has been expected and pre-determined long ago. The mechanism of Gene-suppression-activation is a scientifically recognized and technically feasible procedure, by which a 3,500 million year old master-plan could get accurately timed, and ultimately achieved. And may I reiterate again that all of this is about *how* we came to be.

The *why* is beyond the scope of the scientific human mind. It is for philosophers to ponder, from now till eternity.

> > > > >

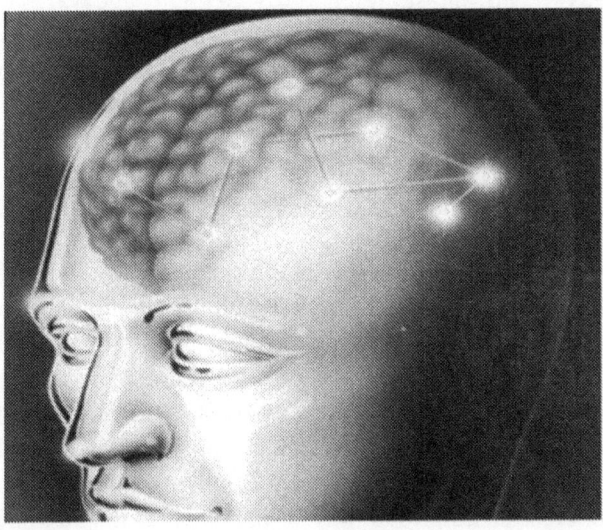

Figure (18)
Even if science would prove that our memory, subjective feelings, and
personality, are all affected by physical reactions of our neurons rather than by
a separate soul, so what?
Mankind is not defined as the animals that have a SOUL but as the animals that
can INVENT ideas and then TALK about them!

There is one important last word about the input that reaches the
human mind. The input to the mind is limited by the limitations of our
five senses. Our eyes see only a certain portion of the wide spectrum of
electromagnetic waves that fill the Universe. Our ears can hear only a
similarly limited range of sound waves. The same holds true for touch,
smell, and taste.

Butterflies can see ultraviolet waves which we do not perceive.
Bats and dolphins can hear ultrasonic waves which we do not hear. A
dog's sense of smell is thousands of times stronger than ours. As described
earlier, Christmas Island crabs have a sense of timing by certain means,
of which we don't have the faintest idea. Birds can cross the wide oceans
guided by magnetic fields of the earth, a sensory input completely beyond
us.

Not only birds, but even tiny butterflies, do migrate. The Monarch
Butterfly regularly makes a two-way migration going south in the fall, and
north in the spring. They make a huge cloud of billions of butterflies flying

about 10,000 feet high. Interestingly enough, billions of bats 'get word' of the butterflies swarm, and simultaneously go up in huge waves to intercept them. Many butterflies get eaten up in each trip, and all this takes place in the darkness of the night sky. This funny drama was recently documented and photographed in video by David Attenborough in the skies of Mexico, using a high altitude balloon. How do these Monarch butterflies know their way high up there? Even more intriguing is how do the bats, from inside their caves, *feel* them, and then take off for their fateful rendezvous?

Another funny story is that of the *smell* of salt. The sodium ion in salt is essential for the function of nerve and muscle fibers in animal life. But sodium is scarce in plant tissue. Vegetarians know that purely plant diets do not have enough sodium. In the wild, many plant-eating birds and animals have been documented actively looking for salt in most bizarre ways. Some birds cling to certain cliffs hammering their beaks against the rocks for hours, to get their supply. Elephants were photographed by infrared cameras pushing their way in dark caves. Inside, they use their tusks to scrape the salt off the cave walls, and then sniff it up their trunks. As far as our mind can tell, the birds and elephants might be guided by a certain smell emanated by salty rocks, but surely it could be something else.

Scientists do not accept as *scientific* anything that cannot be perceived by our senses, either directly or indirectly through scientific experiment, the input of which ultimately passes through our eyes and ears. I don't agree with that. We have to accept that at least a part of the forces of this universe are beyond the comprehension ability of the human senses and mind.

WAS GOD AN ASTRONAUT?

Earlier I promised a story claiming that aliens visited Earth as Gods, and implanted wisdom and intelligence into Mankind about 30,000 years ago. This would give a handy explanation for the sudden emergence of the human kind of intelligence, which is not easily explained on evolutionary grounds. It is remarkable that such falsified stories could get believed by some human audience. Here is the story:

Scientific researchers, as well as philosophers, have long been baffled by the fact that Mankind-type of intelligence is only thirty thousand years old. It does not squarely fit into evolutionary science, and cannot easily fit into the gradualism of Natural Selection. Fanciful explanations have been proposed to solve this paradox. None of them have any scientific feasibility that can be compared to the Gene-Suppression-Activation Theory. One of them gained much publicity among lay people magazines, and writers of science fiction. It says that our intelligence genes were inseminated at the dawn of time, into the wombs of our ancestor mothers, by intelligent space visitors.

It all started by a book titled *"Chariots of the Gods"*, [Von Daniken, 1969], The book was first issued one year following the famous discovery in 1967 of the *'Little Green Men from Space'*, described in Chapter 4. The discovery sparked the imagination of science-fiction writers.

The author points to the story of a human virgin getting pregnant through super-natural means, a story oft-repeated in the legends of many cultures, as well as the scriptures of many religions. He also points to legends vaguely describing fiery vehicles coming from the sky, carrying "Flying Gods", a legend oft-repeated in many cultures: Sumerian, Indian, Eskimo, Red Indians, and Tibetans. He points to ancient drawings in Sumerian and South American civilizations, which vaguely depict men with what looks like astronaut helmets and what looks like antennae, on top of their heads. Of course the resemblance is superficial, and the drawings are primitive. They could be explained otherwise in many ways.

In this book, Von Daniken clearly claims that: the human race is an act of deliberate breeding by unknown beings from outer space. On page 69 of the book he further deliberates as follows:

> "An unknown space-ship discovered our planet. The crew of the space-ship soon found that the earth had all the prerequisites for intelligent life to develop. Obviously the man of those times was no homo sapiens, but something rather different. The space-men artificially fertilized some female members of this species, put them into deep sleep, so ancient legends say, and departed. Thousands of years later the space travelers returned and found scattered specimens of the genus Homo sapiens. They repeated their breeding experiment several times until finally they produced a creature intelligent enough to have the rules of society imparted to it. The people of that age were still barbaric. Because there was a danger that they may regress and mate with animals again, the space travelers destroyed the unsuccessful specimens or took them with them to settle them on other continents. The first communities and the first skills

came into being; rock faces and cave walls were painted, pottery was discovered and the first attempts at architecture made. Those first men had tremendous respect for the space travelers. Because they came from somewhere absolutely unknown and then returned there again, they were the gods to them"

The writer further argues that building the Pyramids and colossal statues in Egypt, South America, and Easter Island, all required architectural expertise which were not available at the time. He also claims that the Pyramids must have been built by visiting intelligent creatures as landmarks for their next visits, and possibly also as storage places for their wisdom.

Waiting for alien space visitors is a cult that finds many believers in the twentieth century, so much so that some of them may get easily lured to suicide. Late in 1997 the approach of the comet Hale-Bopp was taken by 39 Americans as a signal of incoming visitors from space who would take them for a better life among the stars. Each of them packed his suitcase beside his bed, and took enough drugs to guarantee him a dignified suicidal farewell to mother Earth. This incident did not take place in a jungle, or in the African outback, it took place in San Diego, California, USA.

The book also tells of *'copies'* of ancient maps which were *'discovered'* two centuries ago, and kept in the Topkapi Palace in Istanbul. One map looks as if it was an aerial view of the Earth's continents drawn by someone at a very high altitude vertically above the pyramids of Cairo, Egypt. The allure of the book waned over the years. So, in 2003 a new edition was published, this time with a more direct title: *"The Gods were astronauts."*

In 1999, a well-researched reference titled *"Ancient Mysteries"* was published, [James and Thorpe, 1999], its authors, are renowned authorities in the field of archaeological history. They had spent several years of in-depth study of the most popular mysteries of the past. They showed many of them as outright fakes.

The roots of each fake fell under one of two categories: the first albeit good-willed, is insufficient and amateurish research. The second was outright fraud. It consisted of a few facts which get falsely twisted, but sometimes was totally fabricated artifacts. The motive was lots of money to be gained by highly publicized *'book-selling-fraud'*. This refers to ideas that fascinate the imagination of the people, in spite of being untrue. In at least one case, the motive was political: Ian Smith, the racist ruler of Rhodesia in 1965, revived an old story. The false tale claimed that the Phoenicians, rather than black African natives, were the original builders of the famous Zimbabwe Great Towers, which are a tourist attraction for the country.

In December 2002, a certain group calling themselves *The Order of Angels of the Raellian religious cult* suddenly announced that they had cloned the first human baby. The announcement came just before Christmas-time 2002, just in time to fill the holiday news vacuum. Almost everybody believed that they are charlatans. Their prophet, called Claude Vorilhon Rael, claims that a group of short green space aliens visited him 30 years ago in a French volcano, and revealed to him that all of Mankind are descended from clones they planted on Earth only 25,000 years ago. Oddly enough he has 55,000 followers or *'Raellians'* so far.

In every case, the forgers easily exploit a well known weakness of Human nature. We get easily fascinated by the romantic appeal of mysterious events at the time of our ancestors. The hoaxers may believe that nobody is harmed, as science fiction tales, but they are mistaken. The twisted facts linger for decades in the minds of people. As illustrated in the Hale-Bop comet mass suicides mentioned above, twisted forged legends may actually cause terrible harm.

CHAPTER 10

* * * * * * * * *

GENETIC HUMAN NATURE

* * * * * * * * *

In this Chapter, while studying Human Nature in its raw genetic form, we are only looking for the facts as they are. We do not go beyond that into defining *virtue and evil*. Further in that study, it makes no difference how this Human Nature came to be; whether by direct creation of Adam and Eve, or by divinely directed evolution.

THE BASICS, AND UNIVERSALITY, OF HUMAN NATURE

Human Nature is a group of universal in-born genetic traits and emotions that are common to all human races and groups. It is the same for every member of Mankind, not only as grown-ups but also as new-borns. A *NEWSWEEK* article (Aug. 15, 2005) is titled "Reading your baby's mind"; it compiles recently proven scientific evidence: Emotions such as jealousy, empathy, frustration, and the ability to read facial expressions ...etc are all present beyond doubt in very young babies. It further proves the genetic origin of a hard-wired human nature.

If left to their nature, people will do everything that is listed in this chapter. However, nurture and culture do have a great effect, augmenting some traits, diminishing or even suppressing others. Nurture and culture differ from one ethnic human group to another and from one person to another. Cultural factors in this respect may be comparable to the knobs of a factory-made television set. They may increase or diminish the volume and tone of any particular genetic trait.

Some snobbish philosophers, mainly Western, deny that such a thing as a universal Human Nature exists. They claim that the Human Nature in white modern societies differs from the Human Nature in the rest of the world. They call themselves 'postmodernists'. This is the same racist concept that made Hitler believe that Teutonic Germans should rule the world. The more you read 'postmodernist' literature, the more you will realize how mistaken and disillusioned they are. For example, Blacks and Native Americans who get equal opportunities of education are no different from main stream Americans.

The study of individual human nature includes individuals who become kings, rulers, and emperors. They are mere humans, with all the good and evil of any individual. A country ruled by an evil dictator, whose whims are law, may soon look as if it is an *'evil country'*. History abounds with examples of brutal wars initiated by such evil countries or empires.

Before I start discussing this interesting list of genetic traits, one important point needs to be strongly stressed. It is about the three essentials, *food, cover, and shelter*. They are the three most basic of bodily human needs. If someone is denied one or all of them, his vulnerable body will soon perish. Before he perishes he will be nearer to beast than human. None of the next pages will be relevant, or applicable, to him as a human being.

Now, let me list in the following box the in-born genetic traits that have been proven beyond doubt to belong to our species. Following that I will summarize only the pertinent points about each trait.

IN-BORN GENETIC HUMAN TRAITS

Consciousness, Dreams of immortality,
Selfishness, Narcissism, Envy,

Tool-usage, and Long-term planning,

Private ownership, Desire for more (Greed), Vanity and Fashion,
Nakedness, Body display,
Sex,
Sex versus Love, Jealousy, Polygamy, The wedding celebration,
Incest-Avoidance,

Feminism,
Concepts of Work and Money, Charity versus jobs, Menial jobs,
Pride,
Concept of Government, and lust for power,
Dictatorships,

Concept of Free-Will, Religious feelings,
Morality, Altruism,

Mass Psychology...When Mankind changes to *dinosaurs*
Revenge,   Forgetfulness and Forgiveness,
Aggression,   Homicide,

The Social Animal: Feeling-needed, Herd-approval, Tribalism,
Sibling rivalry, Nostalgia, Ethnicity,
Teenage-rebellion, Globalization,
Gossiping, Voyeurism, Privacy,
Facial expression, Boredom,

Sympathy for the helpless,
Natural aversion to snakes,

Beauty, Music, and Art,

Human nature and World peace,
The survival value of Human nature,
The Devil and Human nature. The Devil in Evolution Theory.

> > > > >

CONSCIOUSNESS

A dog owner would bet a thousand dollars, that his dog is as conscious as he, and his children, are. The subject of *human-type* consciousness will be dealt with in the next chapter, titled Body, Mind and Soul.

DREAMS OF IMMORTALITY

If someone would get close enough to any American president in his second term he would discover a very interesting observation. Every one of those presidents becomes obsessed with his place in History. He has won a political-career climax by getting elected for a second term, and is approaching the end of his public life. He starts to think of his symbolic *afterlife*, both while alive and afterwards. This phenomenon is not a monopoly of American presidents; it is one of the most basic characteristics in the Nature of human beings. It is more or less the same, starting from the Oval Office down to the chimney sweep or the bathroom cleaner.

Look into the hieroglyphics of the Egyptian Kings and Queens. It is all about their merits owing them immortality, whether on divine basis or because of heroic and good deeds. Look into the peasant toiling day and night on one meal a day, in order to send his son into school. His son is the symbol of his immortality, and he wants him famous, and glorious, carrying his name. Look into the celibate hermit, praying day and night, having only bread crumbs for food. His heart is fixed on the immortality of the Nirvana in the Afterlife, and nothing will deter him from that aim. Millionaires create trusts to spend on universities, hospitals, libraries, and other philanthropic purposes carrying their names. This would preserve their names much longer than their spoiled children, who may just throw the money into thin air. Alfred Nobel's memory will be renewed every year, forever, by the *ingeniously planned* prize-package he made.

SELFISHNESS VERSUS ALTRUISM

How would you translate the word *mine*? It means that it belongs to my self. Add a suffix -ish, and there it is: selfishness. As a human it is just natural to be selfish, *you have to make an effort to act otherwise*. A society may find it beneficial to help these individual efforts by social norms of law and order, or by adding religious zeal, *but it is NOT natural!*

Out of the instinct of selfishness, neither the ability to thank nor the ability to forgive does come naturally to human beings.

The *only* naturally genetic non-selfish trait is motherly love. It is the only manifestation of pure altruism that gives, and gives again with no expectation of any reward. As to mothers who abandon their children, they are the exceptions that prove the rule.

NARCISSISM

Narcissus, in Greek mythology, was the beautiful son of a God and a Goddess. Vain and aloof, he spurned the love of all the beautiful virgin nymphs, and went to the lake to behold and admire his own image mirrored on the water surface. He was transformed into a flower, the narcissus, which became the symbol of heartless beauty. Narcissism denotes an extreme degree of self-esteem and self-involvement, the ultimate vanity. Look around you, and you can easily see examples: the snobby pampered princess, who ends up as a miserable spinster, the self-admiring executive who leads his company into ruin, Emperor Nero, who believed that he had the ultimate wisdom, etc.

All Dictators, sooner or later, fall under the spell of narcissism, leading to unavoidable detriment to their people. History has proved it repeatedly. In studying history, the present era is often over-emphasized. This is because of the personal impact. But past history is full of even more dramatic examples.

ENVY

The passion of envy is commonly ignited in childhood by the slightest sign of more favor towards a brother, a sister, or a schoolmate. The fire of childhood envy commonly survives, in the subconscious, into adult life, and then ruins the life of its victim. He will always feel less worthy or less privileged than all other humans. The assumption will haunt his every decision, action or behavior. Soon, his originally false belief will make him look as if he is fulfilling an omen!

Bertrand Russell claims that *"The human instinct of envy is the origin of all democratic theories, everybody should be equal. An extreme form of 'equality' is communism. In communist countries everybody is 'equally' poor."* (Emphasis added). This shows how far the human passion of envy can destroy not only a person, but also a whole society. Kings, rulers and dictators are mere humans who harbor the same passions as you and me.

The cure of envy in childhood is mainly preventive. In adulthood the cure is philosophical, by encouraging everybody to derive pleasure from what he has, rather than deriving pain from what the *others* have.

TOOL-USAGE AND LONG TERM PLANNING

Monkeys break open hard nuts by using stones or logs of wood. Otters hold two pieces of stone under their armpits, and use them to crack-open the shells they catch. A chimpanzee cuts a small tree branch, removes its leaves, and then uses it to extricate delicious termites from deep crevices and cracks. I recently watched a beautiful film of wild nature, which shows an Egyptian-Grey-Hawk breaking the egg of a prey bird for its own food. He holds the egg in his beak and raises his head high up and then hurls the egg on the ground to break. If the egg is too big for its beak, he picks a stone, and repeatedly and forcefully hurls it down on the egg until it breaks. Long after the egg has broken open he keeps picking the stone again and again, throwing it down on the egg shell until it is completely shattered into small pieces. This last part of the effort is not necessary for his meal anymore, but it is an automatic gene-imprinted action for which he is using no intelligence at all.

Even more significant than similar examples of tool usage, is the following example of long term planning. A certain subspecies of ants is called leaf-cutter ants or Parasol ants. Thousands of its workers cut up tree leaves in measured sizes that allow each ant to carry each piece alone. Onwards they march in long valiant and astonishing lines into a specially prepared chamber in their huge under-ground nest. The leaves get spread on the floor, and then chewed into minute sponge-like bits. The ants cannot digest the leaf's cellulose; they are just using the leaves as a raw material on which a certain fungus will be cultivated. Tiny workers start spreading thin strands of the fungus on the chewed leaves. Within 24 hours the green color disappears, and the fungus has developed a massive carpet of tiny mushroom blobs. These are the food after which the ants have gone into all this trouble. They start harvesting, and eating, their mushroom crop. One or two days later, they remove the digested leaves into waste heaps outside the nest and clean the chamber. In the next morning they start another agricultural cycle again.

It is worthwhile noticing that Mankind did not have such an *instinctive agricultural behavior* laid into their genes. They had to spend thousands of years using their eyes and minds to observe, deduce, and discover how to plant seeds which later grow into food. This agricultural insight was the first step towards the civilization of Homo Sapiens. In contrast, the complex agricultural system of the Parasol ants is just a genetically inherited behavior. It was neither discovered, nor developed any further, since millions and millions of years.

All these examples of tool-usage and long term planning in the Animal Kingdom have been the same for millions of years, with neither change nor improvement of technique. Chimpanzees have been around for millions of years. Even if they could learn a few tricks from creative mothers, still they had to 'reinvent the wheel' over and over again with every generation. This is in complete contrast to Mankind. He has been around for only a few tens of thousands of years, yet his *cumulative* culture has carried him to incomparable extremes.

PRIVATE OWNERSHIP, THE ETERNAL SCOURGE

Acquisitiveness is a most powerful and exclusively human instinct; there is nothing comparable to it in any other form of life. I believe that our *Original Sin* was this instinct of private property, rather than the sexual adventures of Adam and Eve. Private ownership is one characteristic from which no human is exempt. It is as dominant as the ability to invent, and the ability to talk. Realizing their ultimate mortality, humans have devised complicated rules of inheritance, just to keep their wealth "in the family". In children this instinct starts to appear just at the same time as language, at age two to three. A child of that age will emphatically say that the ball belongs to him, and not to his brother.

Professor Wilson of Harvard University in his book *"Consilience"*, page 171 says that all mammals, including humans, form societies based on a conjunction of selfish interests. Unlike the worker castes of ants and other social insects, they resist committing their bodies and services to the common good. Rather they devote their energies to their own welfare, and that of close kin, [Wilson, 1998].

WANTING MORE, (ALIAS: GREED)

Do you remember your childhood story of King Midas?! He wanted more and more gold; ultimately he changed his own beloved daughter into a statue of gold. The desire for more is the eternal nightmare of everybody, starting from the billionaire, down to the beggar on the street. Often the hair-thin difference between happiness and misery is whether you spend as little as one percent *less*, or *more*, of what you earn. Nobody feels that he has enough, whether money, land, houses, transport means, or even T.V. sets. Owning more is the easiest way to outshine the neighbors or relatives, thus becoming the usual outlet of *vanity.* Quoting from Colin Tudge in his book *"The Engineer in the Garden"*, on page 367 he says *"The desire of extreme wealth continues to be, or is acknowledged to be, the principal driving force of dominant economies."* [Tudge, 1993], (emphasis added).

VANITY, FASHION

Vanity or showing-off is a secondary human characteristic, well rooted in everyone. It may be rudely advertised by arrogant personalities. Less arrogant people will claim modesty, or religion, as a civil excuse for masking their natural tendency to show off. Only the very pure at heart wouldn't *feel* the vanity urge at all. **Fashion** is the current popular custom, style or taste, especially in dress, housing, furniture, cars, social behavior ...etc. As such it is an industry churning billions of dollars every year. It thrives on the eagerness of the well-to-do to lavishly spend just to look different from other people, as if to be a "status symbol". So, and by definition, we find fashions deliberately changing every year, just to keep ahead of the less privileged who desperately try to catch up. The character of the "newly-rich" is often mockingly represented in novels and theaters.

NAKEDNESS

What characterizes us is not only the *nakedness* of our bodies *it is our feeling of it.* No animal feels naked enough to try at least to cover its genital area. However, we invented clothes not only out of modesty to mask our genitals, but also to protect the very vulnerable bodies we have. Mankind is as naked and vulnerable as the little naked moles which keep hiding in their under-ground tunnels, eating roots and tubers. If they come out they would almost immediately perish, either by the simplest change of weather, or at the claws of a predator. They would never ever dare to venture out into the unknown as we do.

The feeling of nakedness is not a socially acquired habit; it has been with us since the dawn of Man. Most archaic clothes were made of animal skin and fur, which do not leave many traces. They do not fossilize like bones, yet many conclusive examples have been unearthed in pale-ontological fossils.

BODY DISPLAY

The word *"exhibitionism"* is defined in the Encyclopedia Britannica as such: *"A serious offence if it is a male behavior. But female exhibitionism is not identified as devious behavior, because there are socially acceptable ways for women to display their bodies."* (Emphasis added).

Some of the most common social ways for a woman to display her body are at the beach and the swimming pool. Another is a formal dinner party. You see all the men well-covered from neck to foot. But look at the ladies' night dresses. Thirty to fifty percent of their skin is bare. The dress covering the rest is cleverly tailored so as to show all the attractive curves. In ultra-conservative societies, you find all this covered by a cover-all-garment, which gets taken off once the doors are closed.

You could *reveal* the feminine figure in two opposite ways. The first is to shorten and open a dress so as to reveal as many square inches of body surface as you feel compatible with your version of decency. The second and opposite way is to *cover* it all, but in so tight a way that the cover almost becomes a second skin, a *bare* one. With jeans and a very tight blouse you have literally revealed *all* your body.

The human male instinctively values his *work*, rather than his appearance, as part of his self-image. In contrast, the human female instinctively values her *beauty* as part of her own self-image, regardless of any external judgment. She will always try to enhance whatever beauty she has, and will always try to improve on it, even by artificial or surgical means. For her, the eyes of other women, or men, are simply *additional mirrors* that are further assuring her of her beauty's worth.

Look into the handbag of any career woman, you will most always find a small mirror. This is an item never found on her executive male colleague. That mirror has stayed there in her handbag throughout her life, since she realized, as a toddler, that the face in the mirror is *hers*. When she looks at her face in that mirror she is not admiring her Ph.D. degree, or executive achievements. She is assessing her self-image. Unlike men, her face and body features are a significant part of her self-image At that moment she is only a woman, which does not mean she is inferior to men at all, it is just her own natural self.

You have to admit that the female human figure is, today, the basis of the Advertisement and Public Relations industry, an industry creating billions of dollars every year. Feminists trying to stop that will have to admit failure. They claim it as a vicious practice demeaning the dignity of women, but it is not. It is simple human nature that can never be changed.

Female body-display can also become an occupation, even without the involvement of sex. A model, an actress, a dancer, etc. can earn money, and the shorter or tighter the dress the more that money could be. On the receiving end, people will pay money to a woman for every possible square inch of female skin that she reveals.

Being sexually desired by a man is flattering for the self-image of any woman, no matter her professional position. This is true even if she never contemplates any sort of response to that desire. Do you remember the film *"The King and I"?* (Played by Deborah Kerr and Yul Brenner), Anna couldn't make herself respond to the King's hidden desire except when he was safely dead. At long last she reverently held his hand and kissed it, wetted with her tears.

SEX IN HUMANS, ARE WE DIFFERENT?
– SEX VERSUS LOVE

In the whole animal kingdom sex is all about producing offspring. A bitch is sexually receptive only at her ovulation time. Her male gets sexually attracted only by the signs which she advertises during that time. This is not true for Mankind. The female human figure, or even just her silhouette, has become the eternal sexual attraction, round the clock, month, and year. The human ovulation time is masked, and sex has become a pleasure-full reward on its own merits, regardless of producing offspring. **The frontal position, face to face, is unique for Mankind**. It is suited just right by the anatomy of the human female pelvis, as well as the attractiveness of the face, lips, and breasts. The most sensuously sensitive female parts are the clitoris and labia. In this position they are automatically and strongly stimulated by the mutual pelvic thrusts; which are just natural and spontaneous. This is in contrast to being artificially stimulated by the mechanistic, emotionless, techniques of pornography. Further, the frontal position enhances the emotional aspect of human sex. Sex in humans is not a purely physical act; it carries in its features many psychological pleasures, besides the physical pleasure of intercourse.

A couple can never '*make*' love; they can only '*feel*' love. What they make is a physical act that could be compared in the back of their minds to an act of urination, hence, the need to mask it by the ambiance, the dimmed lights, the music, the perfume, the modesty, and all the paraphernalia of mood-creation. What does a couple do when they get startled by a phone? They involuntarily cover their bodies! The recent pornographic idolization of physical orgasm is represented by a flood of articles and books, which are naturally best-sellers among the youth. The one I chose to comment on is a book titled *"Ordinary Women, Extraordinary Sex"*, [Scantling, and Browder, 1993.] The authors describe how some women could achieve heights of sexual pleasure that deserved a new term: "sexational"! The essence of the 250 pages can be summarized in no more than one sentence: The human female can achieve that bliss only if she feels loved, respected, and secure. The human male can achieve such bliss only as a reflection from his female. This is the wisdom that any couple can gain from this study. The idolization of physical orgasm is an offshoot of the meaningless bleak picture of human life created by atheism. I've nick-named it as "mutual masturbation", rather than "making love". It is a vain and hopeless campaign to compensate for the lost real pleasure, which is the emotional background of mutual sincere love.

The real pleasure is the degree of warmth in the relationship. It is the psychological and romantic part of it all, The female voluntary '*submission*', her feeling of being '*powerfully possessed*' by someone whom she adores and respects. Similarly also is the male feeling of *powerful possession of the body and soul of a woman, albeit gentle, caring, and discreet.* If sex is practiced with this background of mutual love and respect, it will then give the human body and soul an ecstasy far beyond the physical love per-se, regardless of any dreary circumstances. This explains the genuine happiness entertained by a poor peasant and his wife, even if they just barely live from one meal to another.

Any other sex, paid, loveless, casual, or forced, etc, will give both partners nothing more than a transient and elusive moment of fun. It is almost always followed by mutual detestation and disgust. The background of the whole relationship will always be enmity and loathing, albeit under many masks.

MAN'S *"ACHILLES HEEL"* IN THE SUBJECT OF SEX

There are two complimentary built-in human traits in this regard. Most men are rarely able to *perform*, and never able to *enjoy* sex unless the woman is *genuinely* receptive and willing. [Someone may argue that rapists *can perform* in spite of the unwillingness of the female. But the answer is simple: in spite of their publicity, rapists are a very small minority of **abnormal** men.]

A human female won't give signs of receptivity and willingness unless she feels the *"L.R.S."* three essentials: being **'genuinely'** *Loved, Respected, and Secure*. Perhaps the human genes are made that way in order to secure the protection, (moral and financial), badly needed by the pregnant woman and her baby for at least a future of eighteen years. But there is one more snag; the woman must be able to *voluntarily* respect her man, in order to genuinely dissolve in his embrace. If she looks him down or despises him for any reason then she can never be sincere, even if he lavishes on her all the love, respect and security he can afford.

The description of these two complimentary genetic traits shows them as a very weak spot or "Achilles Heel" for men. It gives the human female a great source of power, by her ability to tempt or allure a man sexually while refusing to satisfy the desire aroused, except occasionally. This ability to tease a man is an art genetically present in the human female since Adam and Eve. It gives her the potential ability to *ensnare* a man, and then let him believe that it is *he* who proposed to her. Marilyn Monroe sings in one of her films saying *"a man chases a girl, until she catches him!"* It also gives her the potential ability to *enslave* a man. History abounds with examples of that, even in the royal houses of emperors and kings.

SEXUAL JEALOUSY

The story of *"Othello and Desdemona"* by Shakespeare was a master's description of this human vice. Sexual jealousy, nicknamed as 'the green fire', can be much more murderous and destructive than red hot flames. The roots of The Trojan Wars were nothing but sexual jealousy.

Desmond Morris gives an interesting view of sexual jealousy on page 139 of his book *'The Human Animal'*, [Morris, 1994]. He says that the roots of sexual jealousy lie in the need to protect the offspring of the mated pair. Because our young are so demanding and because the parental burden is greater for our species than for any other animal, there is strong pressure to provide maximum security for the growing offspring in the family unit. This can best be achieved by ensuring that both parents devote themselves exclusively to their own brood. Any form of sharing threatens this priority. The evolution of intense feelings of sexual jealousy in our species has been one of the basic mechanisms for maintaining this system of maximum parental care.

POLYGAMY

In a recent book titled *"The Myth of Monogamy"*, [Brash and Lipton, 2000], the authors say that like most mammalians and higher apes, the human male is genetically polygamous. But like any genetic characteristic of Mankind, this instinct also can get tuned up or down according to the cultural norms of various societies. In the 20th century, polygamy within marriage, (or legalized concubines), is still prevalent in Africa, Asia and Oceania, as well as in the Mormon society who mainly live in the American state of Utah. Outside marriage, human history overflows with stories of infidelity, both in real life and fiction. In Western societies, polygamy is a punishable criminal offence. But today, especially in the USA, marriage is much less stable, and the best description for the status quo there is "Serial Quasi Monogamy" (!), rather than monogamy.

THE WEDDING CELEBRATION

The wedding celebration is universal, it differs from one society to another, but it is always there, celebrating every marriage that took place on Planet Earth. No father could deny a daughter a wedding party within the maximum cost he could afford. A girl dreams about her wedding gown and wedding party starting from age 9 or 10. This is initiated by a surge of hormones that floods her body at that age. Hormones are created by genes. So we can conclude that the female wedding gown and party are a genetic trait in human females. Maybe it has an important survival value; the publicity involved is one more guarantee to keep the partner around for the needed care of offspring!

A woman of 70 or 80 may happen to forget many things in her long life, but she never forgets even the slightest details of her wedding party (public celebration), or nuptial night (private celebration). In spite of all the feministic humdrum, losing her chastity without these two celebrations is a very cruel memory in any woman's life.

INCEST AVOIDANCE

This is also a part of the sexual architecture of human life, especially between parents, their offspring, sisters and brothers. This general rule was only broken in certain archaic societies, as in some Pharaonic royal dynasties. They believed that their divine royal blood shouldn't be allowed to mix with common blood. Incest avoidance is originally a *genetic trait* present in all sexually producing forms of life, even in the Plant Kingdom. An important discussion of the relationship between this *genetic* trait, and human social and cultural taboos is presented later, (page 237).

FEMINISM

Human motherhood is a full-time, 24 hour job, from day one up to a whole of 18 years. The human infant is the weakest and most helpless offspring in the Animal Kingdom. He needs a full emotional maturity before he can face the world alone. To gain emotional maturity a human infant needs *both* parents until he or she crosses from adolescence to adulthood. A mother instinctively keeps her male partner around, for this purpose, by her charm. By the way, raising kids to become responsible citizens is a very demanding job. The problem is that anybody is allowed to do it. While a plumber or a driver is required to have a license before being allowed to do his job! In contrast, all animal offspring need parental care for only a few days, weeks, or months to attain physical maturity. With this physical maturity they can face the world alone. This is because their behaviors after physical maturity are totally governed by built-in genes, and are not dependent on an intelligent inventive brain that needs long years of nursing, as in Mankind.

Moreover, the human mother-offspring relationship is the most important single source of virtue, morals, and the ability to give rather than take. This is because it the *only* sacrificial and purely undemanding human relationship. Virtue and morality gained through mother's contact are far more dependable and permanent than any social or religious indoctrination in later years. Children denied this relationship often grow into emotionally handicapped men and women. The worst thing that can happen to a schoolboy going home, after having evaded peer pressure, street gangs, sex or drug lures, is to open the door into an empty home, with no mother to listen to his little grievances, fears, and hopes.

Talking about feminism and gender-roles in general, it is scientifically realized that that in 90 percent of men and women, the following differences are genetically based. **This does not mean that the remaining 10 percent are abnormal men or women.** They are the exception that proves the rule. These genetically based differences are:

(a) The first priority in a man's psyche is his *work*. Love, children, etc. etc. all come afterward. Wives should realize that. It is not only the work per se, but also its quality of work, for his own pride, and the income it brings in, and consequently his ability to put bread on the table. Failing as a 'provider' is a terrible humiliation for a man. In contrast the first priority in a woman's psyche is *motherhood*. No earthly achievement can compensate her for that.

(b) Male brains excel in tests and jobs requiring three-dimensional thinking. Female brains excel in tests and jobs requiring linguistic, communicational and social skills. Women should better heed these findings, rather than fighting up-hill in careers they are not genetically suited for.

(c) Managerial and political careers demand an obligatory sacrifice of privacy, and abandoning much personal comfort and personal relationships. Statistics, of Western Civilization women, have shown that very few women find their inner female psyche accepting such a huge sacrifice. Hence, it is natural that we find no more than ten percent of women *desire* such careers. **They are just normal women;** they are the exception that proves the rule. But their mistake is that they think that *all* women should be like that. Staunch feminists get positively enraged by the indifference with which most other women receive their preaching.

(d) Male brains concentrate on facts, while female brains have a sort of sixth sense that picks up not only the facts, but also their emotional nuances and inferences. So, there are lots of differences in the wave-lengths of conversation and communication between the two sexes. Understanding these differences is vital for long-term relationships such as marriage, or in work places.

(e) After three generations, and 50 years of experimentation with feministic extremes, ***home-based-work*** is emerging as the priority compromise for many Western women. Thousands of women are able not only to earn good living, but also to acquire extreme wealth. The idea is catching fire also in Asian countries. It should be further promoted by governments and non-governmental organizations in all developing countries. Third World Nations should better heed the lessons learned the hard way by Western women, rather than start from square one, and repeat the Western mistakes, all over again.

THE CONCEPT OF WORK, AND
THE ORIGIN OF MONEY

Man is a social animal; he needs other people to live with. If forced to live alone his soul would soon disintegrate, and his body would soon perish. Why is that? Imagine a single man stranded on an island in the sea, let us call him Robinson. First, look at his body. No man can fend alone for himself. He may be able to pick fruits from trees as food, but will find it difficult to catch prey alone. He may be able to knit some clothes from plant leaves, but he may have never known how to build a shelter to cover his frail body. Soon he will discover that the presence of other people is essential for his own survival and theirs as well.

Everybody will do something better than anybody else, and everybody will trade his expertise with every other man. Everybody will have to do something; he will have to **'work'**. The man who can knit clothes will make some for himself, and will trade the extra cloth for some corn, grown by another man, and so on. This is bartering, trading goods, or services, without using money. It was the first means by which human beings exchanged the results of their 'work'. With the increasing numbers and complexities of human life, bartering proved too cumbersome. The idea of 'money' soon emerged, and the rest is well known. Money, as such, does not exist in Nature, nor was it created by God, *Mankind created it*. As it exists today, money is just a piece of paper; its value arises from it being honored by other human beings.

Work is a very broad expression. It entails every effort a man, or woman, does to gain Money. He, or she, will then use this money for whatever he, or she, needs. Work can be summarily subdivided into four main categories:

(a) Mental, creative, and executive work,
(b) Manual, muscular, and routine work,
(c) Earning money from the bodily qualities of face, voice, and shape. For example actors, singers, dancers, radio and T.V. personnel, advertisers, air-stewardesses, receptionists, etc.
(d) Earning money from the body itself, as unveiled sex.

A jobless young man is a twisted human soul, he could get manipulated into the most horrible and mean deeds, with the utmost ease. Jobless men are the eternal fuel for terrorist organizations and gangs.

In a human society, the work of a man is part of his self-image and self-esteem. Work is an instinctive necessity in Mankind; it is the basis on which a man can build his dignity and self-respect. These are adjectives which, by definition, are specific to us. They are measured, consciously or sub-consciously, by our share in the total pool of work and produce, of our society.

A man's personal pride is not only related to the amount of his share in the work, but also to the quality of that work. The Architect, painter, writer, composer of music or poetry, all find great pride and satisfaction in their creations, even if it is not immediately translated into big money. A professor or teacher, who loves teaching and is adored and respected by his students, is surely very happy, even if his income is moderate or low. While talking about feminism and gender-roles a few pages ago, we noticed how much the idea of work occupies in the priorities of any man.

CHARITY VERSUS JOBS, AN ETERNALLY CRUEL HUMAN CHOICE

Very few human souls would accept charity without hidden resentment, hatred, or grudge. You may spend a lifetime looking after a poor man, only to find him biting your hand, or stabbing you in the back. A Chinese proverb strongly advises you to give the poor man an angling rod and teach him how to catch his own fish; rather than *humiliate* him by giving him a fish-meal every day for years.

Governments may spend billions of dollars in welfare to the poor, one way or another. This is a vicious circle which never ends, and satisfies nobody. A more creative attitude, based on Human Nature, is to get them busy in menial jobs, even imaginary ones created only for this purpose.

Examples would be planting one million trees on the sides of a road, and when they finish the job they would just shift to another even longer road. Governments can never run short of similar invented projects, by which they would spend the handouts they want to spend; while at the same time satisfying the dignity, and the empty time of the poor people concerned.

The world-wide success of Grameen Banks is a living testimony of how we can exploit Human Nature for the benefit of human beings. The Grameen Banks are banks that give small loans only to the poor and especially poor women; in order to start their own small businesses. The idea was initiated in 1971 by a Bengladeshi economist, Dr. Mohamed Yunis. He dispensed with the costly paper-work and the need for collaterals. Instead, he utilized two of our genetic traits, pride and peer-pressure, in order to guarantee a rate of 98 percent successful repayment of the loans. This means a default of no more than 2 percent, which is far better that what many banks can claim. For more details of Grameen Banks see the article titled *"Banking on the Poor"*, Reader's Digest Magazine, March 1993.

MENIAL JOBS

Do you remember the last wedding party which you attended? Imagine the heap of waste and the mess that had to be cleaned up the next morning. Imagine the amount of stools and urine passed by everybody within the next 24 hours. This is just one of thousand examples of how much filth Mankind would suffer if there were no one to do what we call 'Menial Jobs'. It is a basic necessity for every human group, rather than individual, whether in country or urban life. Surely it isn't a very pleasant job to do, so people must be coerced into doing it, or brain-washed so as to love to do it. The example of coercion is slavery. The example of brain-washing is the 'untouchables' caste of the Hindu caste system. The 21st century version of all this is importing cheap working hands from poorer countries, and then giving them no option but to do the job. Even without coercion or brain-washing, millions of people will just have to take menial jobs as the only work available to them.

PRIDE

Most of the following words and words similar to them are synonymous: pride, dignity, prestige, self-respect, self-esteem, etc., but they have subtle differences. You can have pride in the quality of your work, your painting, or your poem. You can have pride in your good reputation. You can feel proud of the achievements of your son, or your country. But, as a genetic human characteristic that can be easily observed in most of Mankind. I'll concentrate only on two aspects of pride: Pride in the world of work and income; and Pride in the world of emotions and love.

As to pride in the world of work and income, I've just mentioned how humiliating it is for a human soul to be forced to accept charity. Two glaring historical examples are the old feudal systems and the slavery system of white American immigrants four centuries ago. In both cases, the peasants working for the landowners were sure they will have enough food, cover and shelter, as well as protection. But they lacked one thing, and that was dignity and personal pride. Over the years, the resentments of each peasant accumulated to become hatreds, until the collective hatreds turned into mass revolutions. The outcome of these revolutions was dismantling of the feudal systems all over the world, as well as dismantling of the short-lived 'Civilization of the South' in the USA, described in the famous film *"Gone with the Wind"*. Peasants toiling on their own didn't have the security offered by their previous landlords, but they had pride, self-reliance in one's own resources. This is definitely an innate human trait. On the first page of this chapter I counted three essentials, food, cover, and shelter; without which a person can be nearer to beast than human. I can dare add this type of pride as a fourth basic human need.

The pride of self-reliance is clearly evident as well in another field, and that is health. Aging people pride themselves on being able to tend for their own mobility and toilet needs. If someone can't move un-aided, or if he needs toilet care, this will deal a fatal blow to his sense of dignity. As doctors, we see them doing one of two things: either heroically trying to improve the situation, or succumbing to fatal depression. In some Oriental traditions, there is a legend that some wise aging people can actually *will* themselves to die on reaching that stage!

PRIDE IN THE WORLD OF EMOTIONS AND LOVE

A few pages ago, I explained that a human female would never give the sex signals of receptiveness and willingness unless she feels loved, respected, and secure. Without these three essentials, a woman will *always feel humiliated* by the sex act. It is traditional in all civilizations that it is the man who usually proposes marriage, or professes love to a woman, and not the other way around. Staunch feminists criticize these traditions,

and encourage their fellow gals to take the initiative. But alas! Most of
fellow gals are not listening to that advice, why? This is simply because it
is against their innate female human nature. When a man offers presents,
gestures, and sweet talk, he is simply exploiting a purely female genetic
character. Pride, and fear of rejection, will haunt every love affair, from
day one until the end. Each partner will subconsciously look for signals
of acceptance before making any advance. This is an inborn trait in all
humans, and especially so in the human female.

THE CONCEPT OF GOVERNMENT
AND THE LUST FOR POWER

Our symbolic Mr. Robinson was very pleased when other people
joined him on his isolated island. But everybody soon discovered the
necessity of authority to administer even the smallest group of human
beings. They chose some arbitrary one of them to do that job, and gave him
money to be able to carry it out. They also warned him that he will have to
do a good job, or be replaced by another. These are the three basic items in
the concept of "Government": choosing, changing, and giving money, that
is elections, liability, and taxes! This is a basic social contract that Mankind
has genetically found necessary. It defines the relationship between people
in general and the one chosen to administer them.

Government translates to authority, which is the ability to get
things done and the ability to command other people around, it is power.
The word politics simply means the arts, alliances, and intrigues performed
by a person or a group of persons, in order to be given the authority to rule,
or simply to have and be in power.

Yet, no government is more than a couple of meals away from
rioting and revolution. The main duty of a ruler is to supply more food
for the people, more jobs for the young, and more health-care for the sick.
If he fails in any of these duties then he simply must be changed. In the
20th century, and by definition, such changes are obligatory, oft-repeated
humiliations. Anyone taking to politics and public service must accept that.
It is literally a part of the job-description.

The concept of Government is a basic need for any human group.
People feel safer when there is someone at the helm of their ship. If there
is none, they create him, even as a mere figurehead, such as the King of

England, or the Emperor of Japan. This is a psychological necessity for Mankind.

Out of the concept of government we get one further secondary characteristic of human beings, and that is lust for power, status, or dominance, over other people. This demonic characteristic is unmasked in only a small minority of human males, and in even less than a minority of human females. Authority and power in their cases becomes an easier way to fulfill all three of the most evil characters of Mankind, namely: ownership, vanity, and sexual hegemony.

Power creates freedom of action, but often also brings vanity, mercurial mood swings, and constant worry over how long the power will be enjoyed. Reality may merge into fantasy for holders of absolute power; there is nothing to check their will. They seem almost drugged. Few people have the self discipline to handle this drug, a drug which was once described by Henry Kissinger as *"The strongest aphrodisiac in the world"*. A few pages ago it was mentioned how dictators could easily fall under the spell of 'narcissism'.

DICTATORSHIPS

A dictator is an absolute ruler whose word is law. In the past, he was usually the chief of a tribe, the eldest or the wisest man available. In the modern world, a dictator rules a State, much more than just a tribe. A modern State is formed by a mosaic of people with different ethnic, cultural, and religious backgrounds.

Dictatorships can never be purely personal. A dictator can never rule alone, he must have one section of the society as his ruling base. That is to say, it is usually a group dictatorship. In most human societies, there are four main sections: The military, the clergy, the rich, and the poor. Examples of military based dictatorships are all the old Empires, Genghis Khan, Alexander, Rome, The Ottomans, Hitler, and Imperial Japan, etc.

An example of 'dictatorship-of-the-poor' is communism. It started in Russia, and infected many populations before it simply died away. It destroyed whole societies, and in the long run, the persons it harmed most were the poor. The world will need at least two or three generations to recuperate from that fatal disease.

There are two types of 'dictatorship-of-the-rich', an out-dated one, and a modern version. The out-dated one was the old feudal system, in which small elite possessed all the land, and were simply served by the rest of the people. The modern version is what is going on in the USA. Congressmen and presidents cannot get their posts without a lot of campaign donations from the privileged and rich elite. This elite then represents the masked dictatorship of the rich, *they are the king-makers*. The wealthy elite also *control the media*; this further tightens their grip on politics. Further they hold the keys to private donations for scientific research. Hence, all the recent advances of science have simply become profit-oriented. Modern medicines, vaccines, seeds, inventions, etc, have all fallen under what is called patent laws. They have become private property of the research companies, denied to the rest of the world.

Even without donations by others, a wealthy group or person can gain political clout, or even the presidency. This is simply because of his wealth. In the film titled *"Giant"* starring Elizabeth Taylor, Rock Hudson and James Dean, Dean's character has struck oil in his small ranch, and in spite of his lowly vulgar character, is holding a big party attended by all citizens of the town. Amidst a lot of fanfare he is about to be declared Mayor of the town. On entering the hall, he was received by a standing ovation from everybody, even those who deeply resented his vulgarity. They were just carried away by the fanfare. The only un-expected obstacle was that on that particular evening he was too drunk to give his speech. Everybody just left the party, leaving him fast asleep on the bench.

Theocracy is a clergy-based dictatorship. Religious clergy are supposedly too reverent for the humiliations of 'democratic' changes of government. If any of them would accept that, then he is no more clergy, he is a politician in disguise. If a Clergy wants to become involved in politics he should take off his robe, whether a literal robe or a symbolic one.

Pharaohs declared they were Gods. Roman Emperors did the same, Medieval Kings in Europe claimed divine authorization. All this is pure nonsense. Nobody should ever rule other human beings in the name of God. Also no ruler should be above possible change, even if he claims that he is the personal shadow of God.

THE CONCEPT OF FREE WILL, RELIGIOUS FEELINGS, AND "GOOD AND EVIL"

Animals have no choice in how to live, how to feed or how to breed; everything they do is dictated by their genes. You and I are totally different; sometimes we can even defy our genes. We do that every time we use a contraceptive means, and whenever a devout person goes on a hunger strike for a noble cause. This revolt against the dictations of our genes IS our 'Free-Will'. It gives us the unique ability to choose between 'good' and 'evil'.

Philosophers may drown themselves, and their readers, in thousands of pages defining right and wrong. I find this utterly superfluous. All you need is to look yourself in the mirror, alone, and you will almost see, very clearly, the limits of good and evil, as well as see through any slogan or excuse. No one will be able to fool his own self. You may even get your own private definition of evil. Evil is any action that is potentially harmful to somebody else, and you wouldn't like your family or society to know about.

You may even go further than that. The society itself may be a gang of pirates, the society may be permissive in free sex, or the circumstances may get permissive as in mass-rape during wars, but still you will know very well deep in your heart, in front of your symbolic mirror, that in spite of the approval of the others, it is still evil.

Even if I am blind, and cannot look myself in that symbolic mirror, still I can use another eternal wisdom to judge any act. Would I like it if others do the same to me, or to my beloved daughter, son, wife, mother, father, or sister? With that symbolic mirror and this eternal wisdom, nobody would need any books on philosophy, on morals, or even on religion. Most important of all, nobody would need any clergy.

Everybody's version of Right and Wrong is his first step towards human religious feelings. Everybody feels deep in his heart that he will somehow account for what he will say or do. *Nobody* is exempt from this mystic feeling that cannot be explained by any logic or scientific words. The word *nobody* is worthy of bold italics, because this statement does include everybody that has ever lived, or will ever live, on Planet Earth. We will refer to this again under the title of "*THE RELIGIOUS ANIMAL*" later. It is part of our genetic make-up, same as our nakedness, same as our inventiveness, and same as our language syntax.

A book titled *"Good Natured"*, [De Waal, 1996], was dedicated to the research of morality in the societies of apes, monkeys, dogs, and other animals. It claims that our morality and humane behaviors have simply evolved from similar traits present in the whole animal kingdom in varying degrees. In 1964, William Hamilton got very excited when he discovered certain behavior of vampire bats. A bat would die if it fails to suck blood for three successive days. A successful bat would volunteer to regurgitate some blood into the mouth of a less successful comrade, expecting to get a similar favor back in future unlucky days. He called it 'The Hamiltonian Theory of Reciprocal Altruism'. In his view this discovery belittles the altruistic behaviors in Man. [Hamilton, 1964]. None of these observations can be compared to human abstract thoughts, and religiosity.

MASS-PSYCHOLOGY: WHEN MANKIND CHANGES TO 'DINOSAURS'

Why did *allegedly religious* wars outnumber all the other wars in history? There are two main reasons. The first is that religious feelings can easily get deeply entrenched in the hearts and minds of human beings; we are genetically predisposed to that. So much so that many of them could easily prefer to die rather than abandon or betray their particular faith. If this becomes a mass action, then a further factor of mob mentality is added. Mob mentality or mass mentality, is a well known psychological state. It can easily cross the borders of reason to become irrationality. This can take place whether the issue is a religious one, a political one, or even a simple sporting contest, such as a football game.

The evil in human nature can get much more vicious if it becomes mass-evil. This is a dangerous senseless evil in which the human mind takes no share, nor is able to exert any control. An individual throws away all his internal inhibitions when he gets hidden inside a mass of a rowdy mob. This mass reaction is instinctive rather than deliberate. In the day-after participating in a mob behavior, a person may not believe that he himself has shared in such and such actions while drowned in a mob. In a way it reminds me of the pack-instinct in dogs. You may have raised your dog indoors all his life, as an extremely well-behaved pet, only to be astonished one day by this same dog joining the pack in the street, and badly mauling your neighbor's child. The mass action becomes an autonomous living-entity with a life of its own, almost separate from the human beings under its spell. This ominous description is inherent in the human self, even in children.

Most of us have read, or heard of the famous story *"Lord of the Flies"*. It is a masterpiece novel by the Nobel Prize winner William Golding. It tells about a bunch of British boys, ages 5 to 12, who got accidentally stranded on a deserted tropical island. In a few months time they became virtual beasts. They formed warring gangs, two of them were killed, and ultimately they set the whole forest island on fire. The smoke attracted a passing ship, and suddenly they awakened back to what is supposed to be civilized life. The novel beautifully sheds a floodlight upon mass-evil, one of the darkest aspects of the human soul.

In Bertrand Russell's Book titled *"POWER"*, he describes collective excitement as a delicious intoxication, in which sanity, humanity, and even self-preservation are easily forgotten, and in which atrocious massacres and heroic martyrdom are equally possible. On page 103 he warns that un-informed over-confidence of the populace can create bellicose sentiments, which may easily get communicated to the rulers. Hysteria and megalomania are catching, and government people are just human, they have no immunity to such intensive emotions, [Russell, 1988].

A mass of human beings is potentially a very dangerous creature, almost like a dinosaur. Its destructive power could be limitless, but similarly also, its brain-power is as meager as that of the little brains of dinosaurs. Both factors of limitless power and limited brains can doom a human mass in the same way as the dinosaurs were, at one time, doomed.

HUMAN ATROCITIES DURING WARS

Faced with death any minute, Human Nature could soon revert to its beastly form. This is the same as in the discussion of the basics of Human Nature. We warned that a man deprived of the physical basics of life will be nearer to beast than human.

The brutality of ground forces combat was, and will always be, almost a trade-mark of all wars. This includes mass-murder in cold-blood, mass-rape, mass-looting, and a whole list of what a savage intelligent animal can do. I believe that we need not give examples, such as Tartars, Ottomans, Bosnia, Vietnam, Nankin, ... etc, because not a single war in the history of Mankind had missed the honor, or rather the dishonor, of this savagery.

Soldiers who survive such atrocities are often haunted for the rest of their lives by what they did. I dare say to them that such a *private war-crime tribunal* is totally un-justified. They should forgive themselves. The real crime, and the whole blame, is somewhere else, at the luxurious door-steps of the leaders who led them to war.

REVENGE, THE DRAGON OF FIRE AND DOOM

Revenge as a human characteristic is deeply rooted in the history of Man. For some people in rural or nomadic areas it may become the only reason for living. Do you remember *"Moby Dick"*? It is a novel by the American writer Herman Melville, first published in 1851. It beautifully, and symbolically, illustrates an extreme case of the irrational instinct called the lust for revenge.

Moby Dick was also very successful when presented as a film carrying the same name. A pirate called Captain Ahab, played by Gregory Peck, has lost his leg to the bite of a huge white whale which the pirates called Moby Dick. He spent years sailing the oceans looking for revenge. The climax of the tragedy comes when he ultimately meets the whale. He sticks his spears in the heart of the whale, but gets caught in the twisted ropes. When the dying whale struggles away, the whole ship and crew is sunk, seeing their captain hopelessly intertwined with the large beast. Down, both of them go, to die in the deep blackness of the sea.

Like fire, revenge can be incredibly destructive, if left untamed. It could become a legacy, handed over from father to son as part of the inheritance. When a father repeatedly recites tales of old injustices, into the ears of his child, he actually nurtures this instinct in his child's soul. I believe that this is a most selfish and vicious thing to do. He is willfully manipulating the life of another innocent human being, depriving him from the right to choose his own life course. The child will become a captive of his father's past. He will lead his life as a driver whose eyes are eternally fixed on the rear-view mirror; he is doomed, and will go nowhere.

Eight centuries ago Mogul emperors brutally devastated great stretches of Asia and Europe. In one infamous incident in 1223 A.D. they butchered the Russian principalities of Kiev, Rustov and Novgorod. Mongolia today is a backward third world nation just scraping by for its livelihood. No one would dream of chastising a Mongolian peasant or student for what Genghis Khan did, in spite of the fact that he is still a revered idol of national identity in modern time Mongolia.

If human relations are run on hierarchical revenge, religious, personal or political, it would serve no good purpose what so ever. It would neither benefit the old victims, nor punish the real culprits, who would now be very safely dead!

FORGETFULNESS AND FORGIVENESS

For humans, the propensity to forget is a blessing from God. If you see how a family reacts on the day they lose their father you may believe it is the end of the world for them. Have a second look just one year later, and you will recognize the magnitude of this human trait. "Time is the Great Healer" is a correct proverb coined in all cultures since eternity.

A little forgetfulness and forgiveness are also strongly welcome in the relationships between various Nations, if only we could prevent power-lusty politicians or clergy, from cashing-in. Gandhi is quoted to say: "An eye for an eye will make the whole world blind". As between individuals it is also the same among Nations. Forgetfulness and forgiveness are greatly helped by a show of regret, even if symbolic. The whole Far-East was devastated by hideous atrocities that were committed in the first half of this century, by the Imperial Army of Japan. All of this need not go into children's school books, in preparation for a second round in future centuries. A word of regret from a Japanese leader, or a Japanese parliament, would be much more valuable for both sides than billions of Yens in economic aid.

AGGRESSION

Violence is inherent in human nature as a means of resolving conflict. Both aggression and non-aggression are characteristics found in varying degrees in human beings. Generally, aggressiveness is more prevalent in males. At one extreme it can turn its owner into a killer, at lesser degrees he may become the bully of the school yard. When an aggressive person takes to sports he usually chooses rough games, such as boxing, wrestling, rugby, or American football. To earn his living he commonly drifts into the armed forces, secret services, body-guarding, and the like. Aggressiveness could also be limited to the muscles of the tongue. In that case it usually carries its owner into politics. If this is mixed with a lust for power as well, then it becomes a sure recipe for a governor, a company's president, a terrorist leader, or a dictator. Thriller fiction and action films attract people of all ages. They simply seek a discharge or catharsis for aggressive emotions and energy that keeps building up.

Just think of it, in the 10,000 years of recorded human history there has never been a single century without a devastating one or more wars. History books tell our school children about these wars the wrong way. The wars are painted as great victories and conquests of their ancestors. This is a very serious international mistake that must be corrected, (maybe through UNESCO?). The conquest wars of Ancient Egypt, Greece, Rome, Persia, or Tatars, were not laudable empire-building wars. They were nothing but manifestations of aggression, the immortality dreams of their rulers, and the lowly instincts of vanity and wanting more. The National-pride of any country should never be taught as their old conquests, but rather as their share in the advancement of science, and improvement in the quality of human life. Imagine a ten year old into whose mind you implant that the measure of Greatness of a Nation is how to kill other peoples, and bring them as slaves. Old wars should be taught as lessons of 'atrocities by our ancestors', that need never be repeated again.

The advancement of science is making war even more lethal on larger scales. Einstein pointed out the scale of destruction made possible by his discoveries. He is quoted to saying: "I don't know with what weapons World-War-Three will be fought, but surely World-War-Four will be fought by sticks and stones".

Mankind has realized a long time ago that competitive sports could be a very good alternative to real combat or actual war. Just listen to the roar of thousands of people in a stadium, or look at the emotions of millions of watchers on satellite T.V. Imagine if the only outlet for all this adrenalin was through real combat. The Olympic Games were first launched in the year 776 B.C. in ancient Greece. The Greeks found them the best alternative to the endless wars between Athens and its neighboring city States.

HOMICIDE

In the animal world, females fight to protect their babies, and males fight each other for food and sex. But there is always a limit to those fights, they rarely become fatal. No tiger goes out with murder on his mind, to find and kill a fellow tiger, but Man does. In the mind of a murderer there can be a thousand reasons for the act. A child three years old, and addicted to television violence, may find it easy to get rid of his newborn sibling by suffocating him with a pillow. Why is violence so common in fiction, T.V. and cinema? This is because it caters to an unfortunate genetic human trait. But is it a new trend of the 21st century? No! It is as old as humanity. The story of Cain killing his brother, to have his wife, is present in all cultures since the dawn of Mankind. Not a single human culture is free from homicidal stories, or legends, and then it must be genetic.

Tribal chiefs, Kings, Emperors, Prime Ministers, and every ruler, is a mere human being, laden with all the good and bad of human traits. His personal whims can easily get presented as "For-the-good-of-the-Nation", and there you have the seeds of war. The people of Athena and Sparta went into decades of war, for nothing but a jealous King and a promiscuous Queen.

MANKIND, THE SOCIAL ANIMAL

Gathering around a campfire is an enchanting pleasure common to all cultures. In prehistoric times, a fire was the focus of life after dark, offering warmth and safety, fellowship and light. Flickering flames always fascinate our eyes and emotions. This is true whether we are actually participating in a camp-fire gathering, or just reading about it in an imaginary tale, or watching it on the screen, or even just admiring it as a painting hanging on the wall. This sort of pleasure illustrates some of the most basic of human traits, namely: the fear of loneliness, needing and being needed, tribalism or the herd-instinct, sibling rivalry, nostalgia, teenage-rebellion, gossiping and voyeurism, as well as facial expression. Let us have a word about each.

FEELING NEEDED

Feeling needed by someone else, is a basic need of human self. A woman may insist on getting a divorce from a very rich executive. To his amazement he will discover that nothing will change her mind, until at long last he humbly admits that he needs her, he can't go it without her. *A human, (especially women), who feels not needed anymore is a very miserable human soul.* This misery would soon show on the body as well, until he or she actually perishes, body and soul.

HERD APPROVAL

This is one more basic need of any human soul, who lives in a human group? This is the basis out of which kid-gangs get formed, at a very tender age. The gang becomes a separate living entity, almost separate from the individual kids involved. This is another manifestation of the mass-mentality of human groups, mentioned previously.

The best that a mother or father can do is not to isolate their child, but to keep their ties to the kid stronger than, or at least as strong as, the authoritative attraction of his gang. She should never think of shutting him off behind closed doors or protective bars. He will feel detested and rejected by his peers. A rejected member of any society, child or adult, is as miserable as the one whom no one needs, or even worse. More than that he will be deprived of the experimentation of real life, and will rarely reach emotional maturity. Rich and aristocratic clans are abounding with these examples in their offspring.

TRIBALISM, HERD INSTINCT

Tribalism is also a very strong instinctive adjective of human groups. The instinct starts by the close relationships to cousins and uncles, and then extends to encompass the whole clan. It has no direct relationship to the territory occupied by the clan, or to any religious or other beliefs adopted by any member of the group. Territoriality, hence patriotism, religious hatreds, and political maps, are not part of the human nature. The main natural collective tie of any human group is their tribal relationships, which include their language, as well as their ethnic habits, food, music, and folklore.

Within the tribalism human trait, kinship feelings may turn into feelings of superiority over other tribes. Shem's descendants felt superior to their siblings, the descendants of Ham. Hitler felt that his Aryan ancestors were superior to other human tribes. Serbs feel superior to other Slav peoples, etc. To gain credibility, these feelings of superiority sometimes lead to a claim of Divine blessing: *We are the Chosen People.* Well known examples are: Abraham's Jewish descendants, Abraham's Arab descendants, and the Japanese People, Sons of the God of the Bright Sun.

Feelings of tribal superiority could stay as a tribal or national song, and go almost unnoticed for centuries, until a demagogue decides to use it as a trump card. He starts inflaming the notion, igniting a mass mentality craze, which could easily get out of control, and then a Dinosaur is created.

SIBLING RIVALRY

Every mother knows very well that her newborn baby needs active protection from his brothers and sisters. From the child's point of view his parents are his only source of survival. He has to be forced to or lured to, accept a rival who would share their love and care.

This is a simple piece of everyday knowledge today, but look to what it can do to the history of mankind: Queen Victoria reigned for 63 years, she had many children. They became kings and queens of Britain, Germany, and Russia. Kaiser Wilhelm of Germany and Kaiser Nicholas of Russia were cousins, but they nurtured a visceral hatred for each other. Both as well, detested King Edward of Britain, who was their uncle. History tells how this three-some hatred actually contributed to the roots which ultimately led to World War 1 in 1914. Ishmael and Isaac were half-brothers; both competed for the love of their father, Abraham. Several centuries later this fateful sibling rivalry is still alive, and is sometimes viciously fanned, by fundamentalists on both sides, into fruitless wars between Arabs and Jews.

NOSTALGIA

Childhood memories are very dear to every human being. They always occupy a cherished slot in his subconscious self. The father who died in our childhood is always a Saint; the mother is always an Angel. Music was better, morals were better, society was ideal, etc, and everybody cared for each other. It is a sort of idolization of childhood emotions that every human is simply prone to do. Why does it feel great when you meet and chat with an old childhood pal? Because of the shared memories you can enjoy together.

The house of our childhood is always a big mansion, with cozy bedrooms, a rich kitchen, and a beautiful garden if there was one. Faraway immigrants always have a burning desire to revisit their childhood place. When they do, they always get dismayed by the great difference between the romanticized memory and the down-to-earth facts.

ETHNICITY

Besides these manifestations of personal nostalgia, there also another form and that is 'Ethnic nostalgia'. In that context, human nature recognizes only the tribal relations which unite whatever ethnic group. The language, traditions, habits, food specialties, jokes, proverbs, the fairy tales told by grandmothers to grandchildren, the songs, the dances, the folklore, it is all this, and only this, that you miss if you leave your country and settle in America, Canada, Australia or elsewhere. Whenever you feel nostalgia for your original country, you never envisage it as a map, a National Anthem, or a flag. Artificial boundaries created by States are not a part of human nature. They should never get sanctified. Look at the map lines cutting the African Nations apart; they are simply invitations for conflicts and wars.

TEENAGE REBELLION, (*VERSUS WHATEVER PREVAILING TRADITION*)

Human adolescents have a genetic tendency to distance themselves from their families, and to get nearer to their peers. A hidden collective revolt starts to take place, (in the USA, a teenager commonly nicknames his parents as "the fossils"!). Human societies counteract by trying to make them conform to tradition by school discipline, scouting, parades, sports military conscription ...etc. This "tug-of-war" is inherent in every human group. In the hieroglyphs of ancient Egyptians we read about a grandfather wailing over the decadent morals of the new generation. It sounds as if it is an article in a twentieth century newspaper. Similar writings are echoed throughout history in every civilization, down to our own present day. This is in spite of the fact that all the human vices, drunkenness, drugs, adultery, homosexuality, rape, ethnic cleansing, sibling murder, political intrigue, etc, have been with us all the time, even at the supposedly pure times of Prophets and Saints.

Adolescent rebellion, or what is called the "generation-gap", is a universal genetic characteristic. In the long term it is actually one of the forces by which human cultures keep on changing, rather than stagnate throughout history. The genetic peer-pressure for novelty, and risk-taking, almost always, slowly prevails over ancestral traditions. If the same cultural norms were carried over from one generation to another, human societies would have looked really odd today. Nobody knows if this would have been for the better or not. Television, cinema, and lastly the World Wide Web, have transformed the peer-pressure into a global one, rather than keeping it purely local in various countries. I believe that this is one of the most important causes of the often cursed cultural globalization. It ultimately leads to the globalization of everything, from trade to politics to literature. I believe that it was written since eternity, in our genes, that at sometime the whole Earth would simply become a single huge village. (!!)

GOSSIPING AND VOYEURISM, PRIVACY

If a sex scene becomes visible by chance in a neighboring window, it takes a Saint to turn his eyes away. In a west-side sophisticated club or an east-side street café, nobody will be able to close his ears shutting off the latest scandal gossip or joke. The subject of the gossip will always be one of two: sex or financial integrity. Regardless of the degree of truth or fancy, the more famous the person-subject of the gossip is, the more exciting it will be. Reading through ancient history you could almost visualize scenes comparable to both the sophisticated clubs and the street cafés. There is nothing comparable to gossiping and voyeurism in the life of any animal other than Mankind. This type of peeping curiosity is always present in every human being, albeit in varying degrees. In the 20th century, this genetic trait has become the life-line of vast industries, worth billions of dollars every year. They range from tabloid newspapers and magazines, to erotica literature, to various biographies. Biography books occupy one fourth to one third of the list of any publisher, and they are best-sellers. This is because they tickle the readers' voyeurism instinct. Sometimes voyeurism gets lethal, as with President Clinton, Princess Diana, and Marilyn Monroe.

In the classic film *"Roman Holiday"* Audrey Hepburn is a young European princess on a State-Visit to Italy. She gets so bored with the formalities, that she runs away to enjoy a 24 hours as free as any young girl of her age. A journalist, Gregory Peck, shows her around Rome. Unknowingly to her, a photographer takes several shots for her in the streets of Rome. Why is the news agency chief editor ready to pay thousands of dollars for those innocent pictures? Because he knows they will sell his magazine. People love to pry into the private lives of famous celebrities, for nothing but curiosity. Being a part of human nature, it can never be totally erased by any degree of sermonizing or moralization. Before seeking celebrity, fame, or public office, you should better remember that there will always be a price of personal privacy to pay.

FACIAL EXPRESSIONS

Nobody teaches a newborn how to smile. It is a part of his genes, executed by a certain arrangement of the muscles of the face. In all body parts muscles are attached to bones and joints, around sphincters, etc. But in the face they are attached directly to the skin. They are skeletal muscles, so you are supposed to will them to act. But actually, and most of the time, they act spontaneously, being stimulated by your emotions. Sadness, grief, anger, love, anxiety, etc, all get drawn on your face by a clever painter hidden in your mind. The face betrays every thought that goes on behind it.

Blushing is an involuntary reddening of the face in response to certain emotions, such as shame, embarrassment, shyness, and self consciousness. Such emotions are more common in young females. Recently, I heard a very significant comment on the changing norms of urban societies. The comment was that in the good old days, a girl would blush if she gets embarrassed, but today she would get very embarrassed if she is caught blushing, supposedly betraying her innocence.

One big disadvantage of the cosmetic operation known as face-lifting is that a great deal of those muscle attachments to the skin gets lost. The face looks like a wax mask, with very little emotion showing.

Everybody can see the difference between a hearty smile, and a yellow, a business, or a fake one. Even if you are a poker-faced person trying hard to mask your feelings, you can be betrayed by the tiny muscles that dilate the pupils of your eyes, causing them to shine up.

As doctors, we commonly see patients with hemiplegia, who may not be able to will a smile. But when a loved person suddenly opens the door you immediately notice how a *real* smile shines over their paralyzed face. The pathway of an emotional smile passes through routes that are different from the ordinary nerve supply of the muscles.

Tears are continuously secreted by the lachrymal glands to moisten the surface of the eye. Tears overflow if the eyes get irritated. But they *also* overflow as part of many emotions, from sadness to grief to joy. A good actress must be able to churn an emotion violently enough as to start a teary eye that looks natural. Tears as an emotional expression are unique to Mankind. Evolutionists consider tears as an unsolved evolutionary mystery, because they are of no apparent survival value.

Facial expression, including blushing and tears, is another form of social communication, the same as our genetic language syntax. Both are unique to Mankind, and are not found in any other form of life on Earth.

BOREDOM

Animals, birds and fish never get bored of repeating what they are genetically made to do, day and night, for eternity. They move or migrate from one place to another not out of boredom, but to look for food and to mate. Boredom is a trait specific to Mankind, human beings hate monotony. Curiosity and the love for change and adventure are the basis on which all advancement of human culture has been made. You may fancy a certain food delicacy, and search for a specialized cook to make it for you. If he keeps on serving it to you day and night for a month you will get bored. You may even start to hate its smell. Fancy cars and furniture get changed while they are almost brand new, the lady gets used to it in a few weeks time, and then gets bored of it in a few months later. Sheer boredom may be a hidden factor in some cases of marital infidelity. Wealthy people are more prone to boredom. In contrast, the poor rarely get bored, because they are too pre-occupied with their basic necessities: food, clothing and shelter.

People can never keep on learning or working all the time, they have to be amused as well. No one can live for duty alone. People who do that are romantic idealists; at middle-age they usually get suddenly awakened to the lighter side of life and how boring their life was.

Our boredom trait is the basis on which trillion-dollars businesses are based: the business of entertainment. Radio, television, parties, theatre, cinema, circus, tourism, magazines, fiction books, tennis, bridge, poker, chess, are all examples of what people do to fill the vacancy of boring days.

SYMPATHY FOR THE HELPLESS

Nobody, even a brutal Joseph Stalin, could ignore the faint helpless meows of a little kitten stranded on a garden hedge. The visible tangible helplessness of a smiling child can usually protect him from a ruthless serial killer. Playing-dead can often spare the life of a junior wolf from a dominant patriarch of the pack of wolves.

Gandhi was the first world leader to use this tactic of human relations in actual politics. He urged all Indians to use peaceful civil disobedience as a weapon, against the mighty guns of the British Empire. This little thinly-clad philosopher-politician literally disarmed the whole British Army of occupation. They found no excuse to use their guns. They became as helpless as the valiant Indian people demanding the freedom of their homeland.

The tiny little, helpless smile of a new born is actually his only life line. By that smile, he virtually enslaves the mother and father, and could soften the heart of even Al Capone or Genghis Khan. A mother who abandons her newborn on the steps of a Church, or puts him in a box on the river, is just using a simple genetic human trait that is almost sure to save her baby.

NATURAL AVERSION TO SNAKES

Repeated observations have shown that there exists in all primates, including Mankind, a genetic inborn natural aversion to snakes. This is always observed even if a child, primate or human, has never seen a snake before. This aversion is not found in all animals, on the contrary there are many animals and birds that love to attack snakes, and eat them. At present, there is no explanation for this particular characteristic in all primate species as well as Mankind.

Let us remember that serpents abound in mythical and religious stories, and also in dreams. They fascinate the human mind much more than any other animal species. Fantasies of fearful mythical animals, such as dragons, are often made in the shapes of big serpents, with hugely fanged heads that can breathe out flames of fire.

BEAUTY, MUSIC AND THE ARTS

Last but not least, the only trait which is not potentially evil, namely our appreciation of beauty, music, and art. This has long been hailed as the one human character not genetically represented in other animals, the same as language.

This conclusion sounds very badly in the ears of evolutionists. They looked very hard for whatever evidence to prove otherwise, even on the flimsiest of evidence. The only example they claim is that of the bower-bird. After knitting its bower, a male bower-bird decorates its entrance with blueberries, flowers, and any available artifacts. Sometimes it even paints the door margins blue, using blue fruit juices. Then the female comes over, she carefully inspects several bowers, apparently judging the artistic talents of each male, before choosing one of them.

Bees and pollinators are not attracted to various flowers by their appreciation of beauty. They do not choose. The color, shape and scent of every flower is just genetically imprinted into the little brain of its specific pollinator insect, animal or bird. This simply guarantees that the valuable pollen would go to a flower of the same plant.

NATURAL BEAUTY

The human self has a very soft spot for Nature's beauty. Examples are the greenness of the fields, the blueness of the sky and sea, the splashing sounds of ocean waves, and the cool touch of light breeze on your face. Doctors have discovered that natural soothing pheromones get secreted in the human brain when exposed to such beauties of Nature. A twenty minute session of exposure to one of these beauties actually *lowered* the blood pressure of human volunteers who had continuous monitoring apparatus affixed to their arms. Beach and scenery tourism is an industry of billions of dollars. Millions of people earn their living from it. It is all based on this genetic in-born human trait.

Intensive care hospital rooms used to be gloomy well-lit wards, full of bottles, wires and electronic gear. In a recent study of American hospitals it was discovered that a large window overlooking a natural scene will significantly help hasten the recovery of many patients. Now many intensive care units do falsify that effect by fixing large-sized pictures of natural scenery on at least one wall.

Recorded tapes and compact discs are now advertised, giving background sounds of Nature, such as sounds of splashing ocean waves, a faint recording of Niagara Falls, the sounds of a forest with bird songs and animal calls, etc. Another type will cater to those who love country life, it will give them the sounds of a country night, with cricket and frog calls, and even the hissing sounds of tree leaves. People buy and use them, because they do give the comfort which their manufacturers claim.

THE FEMININE FIGURE

To their amazement, doctors have also found that the same effects were also obtained in both male and female volunteers, by decent pictures of feminine human beauty, a smiling face, a flowing hair, a curvy silhouette, or even a single female eye with beautiful eye lashes and wide-open pupil.

No wonder that the female human figure is today the basis of the whole industry of Advertisement and Public Relations. Clumsy nudity has the opposite effect; it creates disgust feelings, and rising of blood pressure. The only piece of feminine clothing which will always survive is the Bikini. This is not out of modesty or morality considerations. It is just because of the surely repulsive or negative effect if it gets discarded.

A related theme that is sure to evoke and excite the imagination of every human is common in all cultures, modern or old. It is about the sensual awakening of a young virgin by a powerful, gentle and considerate male. The male concerned is sometimes human; at other times may even be an imaginary God or beast.

RHYME

The human self has a natural affinity for 'rhyme'. It may be rhymed words, which we call poetry, or rhymed sounds, which we call music, or rhymed movements, which we call dance. This is not a socially acquired habit; it is strongly ingrained in the genes of very young boys, girls, and even newborns. In recent well-controlled research this has been proved beyond any doubt. In the next Chapter there is a discussion of natural endogenous chemicals secreted by brain cells. Their secretion is stimulated by all the external stimulants of beauty, music and art just mentioned. They are called endorphins, and they give you a soothing and calming effect... **they make you feel-good.** Poets, singers, and dancers earn their living (or wealth!), from simply catering to this genetic human need.

In collective singing or dances there are tens, hundreds or thousands singing or moving in perfect harmony and unison. This adds together two genetically built-in pleasures, the enjoyment of rhyme, and the enjoyment of massive social human contact. The appeal is overwhelming, whether through participation, beholding, or even just listening over a radio. They include choral songs, church hymns, group ballets, military parades, folklore dances, among others. Examples have been represented in all cultures, throughout the history of Mankind. Flutes made of bone have been discovered in the pre-historic caves of France and Southern Europe, where cave-paintings have also been discovered. The flutes were in good enough condition to be cleaned and played. Similarly music is present almost everywhere. We find background mood-music in every modern office, shop, or airport, and especially so in every dentist clinic. For every type of place there is a particular type of music to create the specially needed mood, or atmosphere. Disco, rough dance and music belong to the aggression trait in the human self. It is the same catharsis you could obtain by watching or reading a violent-action thriller film or tale.

HUMAN NATURE AND WORLD PEACE

As if they are highly infective, like a virus, the dreams, whims, and desires of rulers soon look to be the desires of the government as well. In no time the virus spreads to the populace in general too. This is a manifestation of one more human trait, *"Hypocrisy"*, for which I found no place to give it a separate heading. Every member of any government on Earth knows very well that he has to sprinkle some hypocrisy on everything he says or does. It takes courage and guts to freely say one's mind, risking leaving a position of authority. Hans Christian Andersen's children story about *"The new clothes of the Emperor"* is a classic masterpiece about hypocrisy. Tsar Nicholas of Russia and Emperor Wilhelm of Germany were cousins, but they nurtured a visceral hatred to each other. The relationship between Germany and Russia was soon painted the same way, and millions of humans perished in successive futile wars.

Wars start in the hearts and minds of chiefs and rulers long before any sword is raised or a bullet shot. Chiefs and rulers are just humans, like you and me, carrying in their genes all the human load of Good & Evil. I've repeatedly mentioned the role certain genetic human traits play in relation to peace between tribes and nations. Just have a second quick look at the headings of this chapter.

THE SURVIVAL VALUE OF HUMAN NATURE

The central core of the Theory of Evolution is what scientists call the 'survival value' of any particular characteristic of any creature. Fast speed, strong muscles, sharp claws, piercing fangs, poisonous secretions, visual acuity, specialized senses, camouflage ability, all are examples of characteristics that help any living organism to out-wit both of its predators and prey in order to survive, and consequently to procreate. As for Mankind, there are only two characteristics that fit this definition: Human inventiveness and the language syntax. Both have enabled Mankind not only to survive, but also to intimidate every other creature on Earth.

But what survival value is there in the characteristics of love of poetry and music, gossiping and voyeurism, vanity and envy, lust for power and selfishness? What survival value is there in the two instincts of private ownership and wanting more? The two last ones have proved to be the most destructive of all the genetic characteristics of Mankind. Both are the main motive for what came next: homicide, aggression, revenge, and last but not least, wars among members of the same species. These same wars are becoming more and more universally lethal, so much so that science fiction stories are now contemplating the end of the human species.

I have promised myself, and you, to discuss only how we came to be, in purely scientific, and physically feasible ways. Ways that need no day-to-day divine intervention.

I have also promised myself, and you, never to ask the dangerous question of why? But thinking of all that evil we carry in our genes, I can't help asking myself: WHY?

THE DEVIL AND HUMAN NATURE

Just look at the list of genetically proven human traits on the second page of this Chapter. Out of about four dozen, there are hardly four or five that can be listed under an adjective other than 'evil'. With that big load of vice, it is a miracle that there are 'good' people at all. I pity every single human having to fight all these evils, just in order to lead a decent life on Earth.

Still, I believe that we have done a good job together, my reader, and me. Do you know what a computer does when you inject it with electricity? First, it wakes up, and then it starts to look at itself, and identify its own load of software. Each of the systems gets its title written on the screen, until the computer has finally 'known' its own-self. Only then it is ready to function in a knowing, rather than an ignorant, way. This is exactly what we have done. We have identified ourselves. We have bared our skin, and then taken everything away, leaving only our naked genes. Knowing our weaknesses, and our strengths, we become better prepared to deal with each other, and with the rest of Mankind.

People naively describe children as innocent. Of course they are not; it is just that they haven't yet been possessed by the whole list of human vices. Babies are bundles of self-interest. At birth, they have already the sadistic ability to enslave mama and daddy, for nothing but a smile, and a smell. It has been proved that a father is attracted to his newborn by a special smell that is genetically inherited from him, and which causes calming endorphins to be secreted in the father's brain. Babies can easily read the emotions of their parents, mainly from their facial expression and mood of talking. Soon, they will master the tactics of how to use their resonant cries to get what they want, when they want. A pampered child can even blackmail his mother by another weapon, for example, by wetting himself, almost intentionally, or maybe subconsciously. Within a year or two, a child can become a positive, and potentially fatal, danger for his newborn sibling.

I believe the Devil is the biggest hoax of all the legends that have been passed down one human generation after another. Mr. Devil is not anywhere, he is inside you and me and everybody. We couldn't make ourselves believe that, and so we gave him a name, and a symbol, and then started to train our curses on his head. Medieval witch-hunting was just another form of this "denial". The witches were symbols of the Devil, who resides in the body of someone else, not in ours. As if by doing that we wash our selves clean. We will have to live with all the evils we harbor in our minds and bodies. Under a thin veneer of civilized behavior, it looks that we are savages in the inside.

THE DEVIL IN EVOLUTION THEORY

When writers refer to immoral characteristics in human beings they usually describe it as 'beastly behavior', or 'the savage animal in Mankind' etc. In his book *"The Human Animal"*, Desmond Morris says that we owe far more to our animal inheritance than we are usually prepared to admit. But instead of being ashamed of our animal nature, we can view it with respect. If we understand it and accept it, we can actually make it work for us, [Morris, 1994].

The inference from all this is that the Devil in human nature originated from our animal ancestry over million years of evolution. This is utterly absurd. Sex may superficially look to be the best candidate in this respect. The scene of a man and woman involved in a sex act looks very much like what animals do. But kindly look back at the chapter "Sex in humans, are we different?" You will see that sex in Mankind is far different. It is overloaded with much more psychological and emotional factors than any zoologist can scientifically ascribe to any animal species.

Imagine two families: a family of a lion, lioness and 3 or 4 cubs, and the other a man, wife and 3 or 4 children. Then look at the genetic human traits list. Even if you over-stretch your imagination you can hardly find four or five that could have an animal ancestry. The lion family is governed by lion genes, which tell them how to look for the next meal, how to procreate, and how to protect the offspring for a limited span of time. Beyond that their genes have nothing more to tell. The lion family lives in the 'here and now', they live in the present. No lion spends hours regretting a financial loss in the long past. No lion gets a sleepless night brooding over a future fluctuation in the stock market, or its own future mortality. The past and future are two of the worst curses that haunt every human mind. But worst of all is the curse of 'private ownership'. Every human will do all he can do to increase his wealth during his lifetime, just in order to let it be inherited by his children.

What survival value could evolution intend by creating the sinister traits of private ownership, vanity, and wanting-more? The three of them are at the root of almost every human conflict, whether on the personal or world levels, even those that are falsely painted as 'religious'. The three of them could potentially end in the very demise of the human race!

DR. JEKYLL AND MR. HYDE...
AND NOVEL WRITING

Do you remember the story of Dr. Jekyll and Mr. Hyde? It was originally written in 1886 by a Scottish writer, Robert Stevenson. The tale is narrated by a lawyer, Mr. Utterson, who experiences first the terror of meeting Hyde, and then the shock of realizing that Hyde, a murderer, and Dr. Jekyll are one and the same man. The story has become an almost international folklore tale in every culture. Why? Because the story actually describes a fact: There are both Jekylls and Hydes in every one of us, the white and black faces of a same coin. Dr. Jekyll's evils are inhibited by religious feelings and social norms. When he drinks the potion, all it does is to release those inhibitions, and voila: he becomes Mr. Hyde, an un-inhibited evil human. Alcohol does the same, but on a lesser scale.

Another writer, Jonathan Swift, wrote in 1726 a similar famous classic: *"Gulliver's Travels"*. In one episode of Gulliver's wanderings, he spent sometime with a group of horses that could talk. He was away from all human contact. Later, in front of a court he described that period in the following words: "Without other people I found I've lost all my vices. There is no one to envy, to lust for, to steal from, to flatter or brag to, or lie to. I felt a great peace of mind. I even threw away the five large diamonds which I dug out a few weeks earlier. They could have made me rich in this vice-full world." Afterward a member of the court indignantly accused him of insulting the human race. He answered: "No, I see myself what I truly am."

A satire is a literary work holding up human vices and follies to ridicule or scorn. William Shakespeare was very clever at writing satire. He had the intuition of a psychologist who knew very well the innermost feelings and emotions of the human self. Moreover his mastery of the English language helped him put that on paper, in attractive poetry and prose. Bertrand Russell had the same proficiency, but instead of writing novels he presented his experience in a non-fiction way.

One particularly successful and common theme in novel writing is that which describes an episode of looming death. Examples are a wrecked ship, a falling airplane, a stuck elevator, getting lost in desert or sea. In all these situations humans start pondering their lives, and the meaning of it all. **For a brief period of time, they actually discover both God and the Devil, lurking in their own veins.**

I can confidently state that when Mankind suddenly appeared on Planet Earth 30,000 or 40,000 years ago, he possessed not only the famous triad of 'Inventiveness, Language, and a Naked vulnerable body', but also a fourth one: the Devil, eternally hardwired into his genes!

On the other hand, I find on my desk many scribbled notes, preparing for the next chapter. The title is BODY, MIND, and SOUL. Many of these notes are describing Mankind as the 'Religious Animal', in contrast to all the 'evil' we have just discussed. After reading thousands of pages, I've come to the conclusion that *'Religiosity'* is another hereditary built-in genetic trait of human beings. This not a philosophical speculation, I am describing scientific research; an idea best summarized by Andre' Malraux when he described *"Human Free-Will"* by saying: *"We have refused to do what the beast within us willed to do"*. [Boorstin, 1998, page 244]

DNA is an organic substance called Desoxyribo-Nucleic-Acid. The 'gene' is an information code, expressed in the form of a certain sequence of nucleotides along the DNA chain. DNA is your computer chip, the material hardware. But the gene is the intelligent information that you put on that

chip. **They are not synonymous.** *Genes are encoded information, quite distinct from the chemical medium on which the information is recorded. It is the same as the information conveyed in a book, being quite different from the ink and paper on which it is printed.*

In the encoded genes of Mankind we find fingerprints of both God and the Devil. This is regardless how these genes came to be: by random happenstance, or by a divinely master-minded 15 billion years old plan.

CHAPTER 11

* * * * * * * * *

Body, Mind and Soul

* * * * * * * * *

"To call religion instinctive means that its sources are deeper than ordinary habit and are in fact hereditary, urged into birth through biases in mental development encoded in the genes."

Edward O. Wilson
Harvard University Research Professor

"The Big Bang simply means that at a certain moment GOD created the particles out of which matter, and life, will later form, and at the same time gave those particles the rules with which they will play the game. Our bodies and minds are the direct outcome of that game."

Hussein A. Amin
Professor of Surgery

The human mind and consciousness,
Are they *physical*, or *supernatural*?

First, I must reiterate the definition of the word 'Human' previously described. Mankind is defined not as the animal that has a 'Soul', nor as the animal with a big brain, but as the animal who can invent ideas, and then talk about them: Inventiveness, and Language.

To answer the above question, hundreds of books have been written. Let me list some of them which I've read and scrutinized. Then let me summarize for you the argument that consciousness is purely "physical".

"The Mind-Body Problem", by R. Warner, Blackwell, U.K. 1995.
"The Private Life of the Brain", by Susan Greenfield, Penguin books, London, 2000.
"The Fifth Miracle", by Paul Davies, Simon & Schuster, New York, 1999.
"The Emperor's New Mind", by Roger Penrose, Vintage Press, London, 1988.
"The Arrow of Time", by P. Conveney & R. Highfield, Harper Collins, London, 1991.
"Phantoms in the Brain", by V. Ramachandran & S. Blakeslee, Morrow & Co., New York, 1998.
"Consilience", by Edward O. Wilson, Alfred Knopf, New York, 1998.
"The Astonishing Hypothesis", by Francis Crick, Simon & Schuster, U.K., 1994.
"The Origins of Virtue," by Matt Ridley, Penguin Books, New York, 1997.
"Descartes' Error", by Antonio Damasio, Paper Mac, London, 1994.
"How Brains Think", by William Calvin, Harper Collins, New York, 1996.
"Consciousness Explained", by Daniel Dennett, Penguin Books, England, 1991
"The Hidden Mind", update by several authors,
    Scientific American Magazine, Special Edition, August 2002.

Figure (19),
Even if science would prove that our memory, subjective feelings, and personality, are all effected by physical reactions of our 100 billion neurons, rather than by a 'Separate Soul', so what? This is a pure technicality with no further significance. Mankind is not defined as the animal that has a soul, but as the animal that can *invent* ideas, and then *talk* about them.

The brain of a newborn has already its permanent set of 100 billion neurons, as well as 1,000 billion glial cells, which help support and nourish the neurons. Neurons start with only a few inter-connections needed for the essentials of the baby's life, such as respiration, blood pressure, and gastrointestinal movements.

With the growth of the child, his neurons do not grow in number, just maybe a little in size. They merely get more and more inter-neuronal connections, which record every day's experiences and memories. Sleep is a most important requirement for the nightly archiving and stabilization of all the inter-connections made during the day. Dreams are simply brain's nightly house-cleaning, to make room for new memories.

These inter-connections are called synapses. Messages cross them to-and-fro with the help of essential brain chemicals, namely: Serotonin, Dopamine, Noradrenalin, Histamine, Acetylcholine, etc. The brain is protected from any other chemicals that circulate in the blood, by what is called the '*blood-brain-barrier*'. It consists of tightly packed glial cells that line the walls of the blood vessels in the brain. They prevent the easy traffic of substances from blood to brain, and vice versa. The only molecules that can usually access the brain are relatively small, and highly soluble in fat, namely lipids.

No single human brain is exactly the same as another, even in identical twins. Although gross appearance may look similar, the difference lies in thousands of billions of inter-connections, which record the purely personal and subjective feelings, experiences, memories and emotions. The factors of heredity also shape how a child interprets his surroundings, which in turn shapes the very structure of his brain. All of this ends up in what we call the mind of that person: "a '*personalized*' brain".

Suppose that a King, let us call him King Edward, develops a brain tumor, and the Court surgeons decide on replacing the royal brain by a transplant. Theoretically speaking, and according to the latest science fiction, this could be a feasible operation. The King will be able to live, in an organic sense. *But*, he will not be King Edward anymore, because His Majesty has been removed! His Majesty *IS* that lump of tissue on the nurse's surgical tray!

If an unfortunate child is isolated from other humans in a semi-closed environment, then he will develop very few neuronal inter-connections. Not only that, but also many of his neurons will simply atrophy and get lost, if he does not use the brain neurons, he loses them. This is the origin of a common saying: *'use it or lose it'*. The brain of such a child will hardly support him well in a future human society.

Moreover, it has been proven that there are certain windows or opportunities for learning specific skills at certain specific ages. For example, the propensity to learn music is at about age nine. Foreign, or second, languages are best taught around six years of age for girls. For boys it is one or two years later, at 7 or 8. These windows will last for no more than one or two years, after which the window closes. The genetic wiring that allows that 'window' or 'propensity' loses its neuronal readiness if not used. Learning a second language after that will be an up-hill struggle, as compared with the ease during that lost chance. The discovery and proof of these neuronal windows or fleeting chances had led educational authorities to reconsider the different ages at which many subjects are taught in schools.

On reaching adulthood, every person begins to regret certain incidents in the past, and anticipating, worrying, and feeling anxiety about the future. A young child hasn't developed that heavy baggage of the mind, yet. He can concentrate on, and enjoy the *here and now,* the food, the warmth, and the hugs. *He enjoys the present. He is happy.*

In contrast to this self-absorbed happy infant, let me add some words about happiness in later life. In adulthood, happiness, same as love, is an emotion that can rarely be enjoyed alone, as infants can do. It is much more enjoyable when it is shared by some other human soul or souls. It takes two, or more, to enjoy a joke, a smile, a good meal, a walk on the beach, watching a magnificent sunset or a beautiful view, or even just enjoying good health and good company. Actually, happiness is the ability to enjoy together the small things around you, because life is a big collection of small things.

THE EFFECT OF DRUGS

Anesthetics can cross the blood brain barrier. They stop the activity of all inter-neuronal connections, which means they stop all 'mind' activity, and consequently you lose consciousness. It is the activity of the inter-neuronal connections that imparts to you the invaluable trait of consciousness. Other drugs that can cross that barrier are Alcohol, Fatsolvent molecules, Cannabis, Morphine, Heroin, Cocaine, L.S.D., Ecstasy, Amphetamine, Nicotine, and Theo bromine. The last one is found in tea, coffee, chocolate and chocolate ice cream, (specially mentioned because it combines the famous triad of chocolate, sugars and fats, resorted to by affluent people who feel depressed).

A common denominator in all these drugs is their effect on inter-neuronal connections. It ranges from mild to severe stimulation or damping, according to the drug and its dose. A common result is diminishing the heavy baggage of the mind, which is regretting the past, and anxiety about the future. The mind becomes nearer to a child's mind, capable of enjoying the present. The problem with these drugs is their bad habit of gradually losing their effect, demanding bigger and bigger doses, and ending in fatal addiction. In the case of the triad of chocolate-sugars-fats, the addiction is also fatal, but indirectly through gross obesity and diabetes.

FEELING-GOOD, THE BRAIN'S REWARD SYSTEM

Now, a bit of good news: Brain cells do manufacture some drugs with similar effects. We call them *endogenous*, (which means locally made), **Endorphins**; and they are usually associated with a flood of the neurotransmitter called **Dopamine**. This is secreted by certain cell groups on the under surface of the brain. They cause a sense of well-being, they make you feel good. It was discovered that their production can be enhanced by many factors such as: physical activity, tender physical touch such as hugging your baby, warm inter-human relations, altruistic actions, a camp-fire gathering, a good shared meal, a good well-earned deep sleep, reading a story or seeing a movie, laughter for any reason, admiring natural beauty, sex, music, good smells such as perfumes and incense, religious rituals, rhythmic movements as in the some 'Sufi' rituals, and also during meditation. Meditation acts by blocking all sensations, as well as all emotions, including regrets and anxieties.

Endorphins, and Dopamine, that cause a sense of well-being, are not the only response for the above-mentioned relaxing situations. Now we know that another response that floods the whole body is a flush of a chemical called **Nitric Oxide**. It counteracts the physiological effects of stress hormones, as well as relaxes the cardio-vascular system in general, hence lowering the blood pressure. It is the same chemical that causes night erections during a relaxed sleep.

There are other situations that can trap you as well in the ultimate present, through the production of the exciting Noradrenalin, rather than the soothing Endorphins. They can also release you temporarily from your personalized past and future. Examples are racing, gambling, heavy sports, tribal dances, pop dance; Disneyland games and swings... etc.

THE EFFECT OF INJURY

If someone strongly knocks his head, a shockwave will brutally shake the jelly-like brain inside. The neurons will immediately shut-off all their sending and receiving functions, but will stay alive. In consequence, our patient will lose consciousness, until the neuronal activity is regained again, in minutes, hours or days according to the severity of the blow. When a boxer wins a game by knock-out, this is exactly what he does to his opponent.

THE EFFECT OF ANOXIA

Anoxia means lack of oxygen. The neurons are the most metabolically active cells in the body. They demand a continuous blood supply, rich in oxygen and glucose, otherwise they perish.

Let us suppose that Queen Catherine of Aragon becomes very jealous of King Henry's romance with Anne Boleyn. She sneaks to his bed and strongly presses her thumbs against the carotid arteries in his neck for four minutes. The Royal neurons will die. His Majesty will lose consciousness, but will still be able to breathe. If she persists for two more minutes, the breathing neurons in the base of the brain will also die, and so will the King. In the 20th century his life could have been saved if someone immediately puts a tube down his larynx, and gives him artificial breathing. But still he would remain unconscious. Brain cells can rarely recover after anoxia which lasts more than four minutes. If ever they recover, it would take several days, weeks, or never. Fiction and films are fond of that situation. They tell how a devoted beloved person such as Anne Boleyn could keep whispering into his Majesty's ears for hours every day, hoping to re-awaken his neuronal inter-connections. In films, it usually ends by a blinking eye and a hug. But in real life, it almost always becomes a serious dilemma for everybody, family, hospital and insurance company. When, and who would dare to turn off the life-sustaining equipment?

THE PHYSICAL CONCEPT OF CONSCIOUSNESS

One Important Deduction from all these descriptions is that your mind and consciousness are formed, and your emotions are effected, by purely physical and chemical inter-actions, with no supernatural soul needed. The W.W.W., (World Wide Web), is almost like a huge conscious brain, simply because its data are simultaneously revving on in billions of computers and telephone lines round the world. Let us imagine an international agreement to shut off every single computer in every corner of the Earth at a certain moment, then there will be no WWW, the internet will be as if it has been given anesthesia (!). Such a situation can be easily compared to the billions of brain cells simultaneously revving up when you are awake. One day, science will be able to understand the physical riddle of consciousness.

The other side of the argument however says *NO!* The physical and chemical interactions are similar to the electricity passing through computer chips, and they can get blocked or enhanced. But the wisdom is comprehended by an external supernatural force, called 'Soul', which incidentally is immortal. It is here that we, as well as any scientific argument, should stop. This is because there is no way to prove or disprove such a claim.

METHODS OF INVESTIGATION

The body and brain of anybody are observable to third parties. The mind though, is observable only to its owner. It is a private, hidden, internal, unequivocally subjective entity. Neither sensations such as pain, touch and vision, nor emotions such as anger, hate, love, pleasure, affect the observable brain. These sensations and emotions are invisible when examining the brain's external or microscopic appearance.

Modern methods of investigating the brain are PET (positron emission tomography), and fMRI (functional magnetic resonance imaging). They can only show which part of the brain is working at any certain moment, but cannot comprehend the content or messages. We will have to rely on what the patient says he feels. It is hard to see how we could have a privileged *first-person* access to someone else's subjective state.

METHODS OF TREATMENT

There are certain diseases of the mind such as depression, anxiety, neuroses, schizophrenia, Parkinsonism, which make the physical body vulnerable. You may have heard all these names, but you may not have realized that in many of these diseases another system of the body also is diseased, it is the immunity system. Depression or the anxiety the night before an exam can make you much more prone to any passing infection.

It was discovered that many of those diseases of the mind are caused by deficiencies or excesses of the essential brain chemicals mentioned above. Hence, the old way of treatment by lying down on the couch of a psycho-analyst, and talking for hours, is giving way to treatment by drugs that enhance or replace the deficiencies. The most common drug treatments are Prozac, Lithium, and an ever increasing list. Each one has its special indications, and special dangers.

Modern research has pinpointed a common reason for many of our psychological diseases. It is living under continuous and chronic fear and apprehension. Unfortunately, this is the characteristic of urban life in the 21st century, all around the world. Let me explain:

If your hand accidentally touches a hot kettle, your arm will get jerked away before you think. This is an automatic spinal cord reflex that is built-in in our bodies for safety. A similar reflex is in-built in the brain; it is called *'The fight-or-flight'* reaction. Whenever you are faced with a roaring lion, or any life-endangering situation, your body will be gripped with fear. Immediately, and *before* you think, your blood stream will get flooded by three stress hormones. They are adrenaline, noradrenaline and cortisol. The automatic reflex is initiated by a center in the base of the brain called the amygdala. The three stress hormones will shut down all non-emergency services, such as digestion and immunity, and will direct the body's resources for either fighting or fleeing. The heart pounds, the lungs pump, and the muscles get an energizing blast of glucose. Once the life-endangering situation is over, all this will settle down again.

But if fear and apprehension are low-grade and constant, then the amygdala will virtually hijack the rest of the brain. There will be a constant raised level of stress hormones. They will interrupt sleep, impair immunity, and exacerbate everything from acne to ulcers. Anxious people will drink and eat more, will have more accidents, and will be more prone to headaches, muscle aches, stomach-aches, infections, cancers, diabetes and heart attacks. Children can similarly suffer too, and here the damage will be double, because it will also affect their healthy development of emotional maturity.

It is interesting that the fight-or-flight chain of reactions can also be initiated by anger. That is why it is wise not to act under the spell of sudden anger. The same as the reflex of sudden fear, the brain is not in full reasoning gear. Frustration is chronic low-grade anger and it has the same effect as chronic fear or anxiety.

Treatment must start by identifying the causes of chronic fear, anxiety or anger. Next is trying to stop those causes. Third is trying to counteract the stress hormones by stimulating the calming endogenous endorphins of your own brain. Physical exercise is the first well-known tool, but the trick is not only to start, but also to persist.

> > > > >

THE RELIGIOUS ANIMAL
WHICH CAME FIRST? MORALS OR RELIGIONS?

Religious feelings form an important part of human instincts. Thus, they can be easily exploited by either of two extremes. One extreme is fundamentalist clergy subduing the masses, mainly for political power. The other extreme is the profuse literature on atheistic reasoning, which is very appealing to the people in the street. If it succeeds, it surely removes a very important pillar, from an already very frail human psychology.

The reversed relationship between morals and Religions may be a surprising idea for many people, but it is true. It used to be said that Religions are 'useful' because they form the moral reference for Mankind. NO, IT IS THE OTHER WAY ROUND. The moral codes of Mankind came *before* any Religion appeared. We are carrying the seeds of religiosity inside every cell in our bodies. J. Watson's says in his book [*DNA*], page 430 that Atheists do not live in a moral vacuum. They have recourse to an innate moral intuition long ago shaped by natural selection promoting social cohesion in groups of our ancestors, [Watson, 2004].

In early 2000 there were repeated incidents of small children committing serious crimes, including the shooting of classmates and teachers. Heated arguments ensued, both on the public level, and in scientific academia. A normal child should be severely horrified on seeing a bleeding injured person die. Now these children are continuously exposed to T.V. violence, commonly without parental guidance, because mama is at work, or just recuperating from the stress of the day's work. Does this disturb their ability to distinguish between reality and fantasy? Does it get them desensitized to violence? Does it kill their senses in this regard?

A second heated question is about how much these small children know about *right and wrong.* S. Begley & C. Kalb compiled an investigation of these questions in a Newsweek article dated March 13, 2000. For the purpose, they tapped several scientific sources. The question was not only *when*, at which age, do these kids get a moral sense, but also *how* they get it. The following are interesting points from that article:

EMPATHY as a key emotion supporting a sense of right and wrong emerges early and, it seems, naturally. Babies cry in response to the wails of other babies. The response is selective to human cries rather than to other aversive sounds. It is as if the baby says to itself: here is a fellow in distress; I must augment his cry for help. Somehow, there is a built-in capacity to respond to the needs of others. Babies as young as one year try to console others in distress. Toddlers offer their security blanket to a teary-eyed parent, or a favorite toy to a distraught sibling, as if understanding that the very object that brings them comfort will do the same to another. If a sad child is denied a hug, if his loneliness is met with continued abandonment, then he is in danger of losing his natural empathy. Kids who, as 14 months old, exhibit high levels of empathy typically become less empathetic after only six months, if they live in homes filled with conflict, and if they seldom feel a mother's love.

Young children also grasp and believe in abstract ideas such as fairness and reciprocity. When asked, as part of an experiment, how to distribute a pile of toys or a box of cookies to a group of children, many respond with explanations such as "We should all get the same". Moreover children seem to have in-built hypocrisy detectors: "But Mommy, if it is wrong to lie, why did you tell her she looked beautiful?!". In very young age the sense of right and wrong is born of the fear of parental disapproval or disappointment. In middle childhood, seeking approval starts to shift from parents to peers, in form of "social approval".

Empathy, feelings of guilt and shame, ideas about fairness and reciprocity, built-in hypocrisy detectors, etc.; all this, shown at such a tender age by scientific experiment, simply point to the roots of morality being born at the same moment a human child is born. ***The evident conclusion is that they are Genetic.*** .

MORALITY IN ATHEISM

Atheistic reasoning claims that a *matter-of-fact* morality can be created. In other words, it becomes a non-religious morality, without even having to ponder about the relationship of morals to human nature or genes. Morality is a mere necessity, or convention, created just for survival, same as the rules on board a lonely small ship on high seas. Everybody complies with the rules, because otherwise everybody will drown.

As to the self-satisfying aspect of morality, atheists define it as the immortality caused by integrity and good reputation, remembered into future generations. This remembrance may be just in the family, or in the whole society, or even at an international level, according to the scale of the reputation. Atheists also claim that a person with integrity will get his present-life self-satisfaction from his ability to control his beastly evil desires. By these two definitions atheistic reasoning completely dispenses with the duty of having to obey a supernatural deity, or waiting for rewards in a supernatural after-life.

Freemasonry preaches a similar sort of matter-of-fact morality, namely doing well for the mere sake of goodness. But still Masons do not trust the individual's goodwill unaided. To start with, every new member has to vow an oath in a highly charged ceremony. Then he has to regularly attend ritualized group meetings, which simply act as re-enforcement, follow-up, and supervision. Many of the founding fathers and presidents of the USA were Masons.

Ultra-fundamentalist clergy, of every creed, claim the extreme opposite, they claim to be *"God's Agents"*, and that man is too weak against his evil instincts or desires, and must be forced to behave. And of course the clergy should be the coercing force. Their declared aim is to guarantee the *salvation* of Man's soul. This is pure hypocrisy. The clergy's only real aim is political power. Even God's messengers and Prophets, in whose name the clergy claim to speak, have been denied such enforcing authority.

Moreover, their claim simply contradicts the basic concept of all Religions, which is free-will, and consequent after-life punishment or reward.

Historically, the religious instinct of Mankind has often been exploited by several evils. The first of these is demagoguery by dictators, who tickle the sensitive religious nerves of their peoples, in order to stay in power. The second is power-thirsty fundamentalist clergy, which is not much different from the first. The third evil are the charismatic inventors of various cults, mentioned previously. Fourth, are certain charities which work under the disguise of religion. Their only aim is your money, and more than often, the money goes to non-charity diversions.

The last exploiting evil is the quackery of mixing religion with supposedly mystic powers of some humans, animals or birds. Examples are astrology, horoscopy, mind-reading, telepathy, palmistry, psycho kinesis, phrenology, superstition, shamans and faith-healers, psychics, clairvoyants, communicators with souls of dead people and the like. In short, anything that implies sanctity or occult power to any human, animal, bird or artifact. Let me quote about 'Shamanism' from the advertisement of one of their books in a library catalog:

> "Shamans possess healing powers, communicate with the dead
> and the world beyond, and influence the weather and movements
> of hunting animals"

Those charlatans actually hijack peoples' genetic religiosity into their own channels, for nothing but monetary gain, and sometimes sex, if their victims are sufficiently gullible. There is a hair-thin difference between imbecile gullibility, being easily deceived, and stubborn skepticism, the inability to accept reason when it is reasonable.

Without going into any detail, **FAITH** is defined as the belief that Mankind are *somehow* accountable for what they say and do. This should never get mixed with what those charlatans advertise, because it is an insult to both the principle of Faith and Religion, as well as to the dignity of Man.

CONSILIENCE

There are hundreds of recent books talking about morality and faith, and their relationship to modern science. Of the scores of books which I scrutinized, I admired most a book titled *"Consilience"* [Wilson,

1998], written by the famous Professor Edward O. Wilson of Harvard University. Professor Wilson writes out of vast experience in Biology and Human Nature. Talking about morals, faith and atheism, he very cleverly, highlights both sides of the coin in an un-biased way. Here are some interesting points related to the view that Mankind does inherit moral instincts in their genes. Following that I'll elaborate on the particular point of 'incest-avoidance'. This line of argument is so glaring, and also so simple as to get easily comprehended by the most sophisticated mind, as well as the peasant in the field.

Professor Wilson writes that People are innate romantics, they desperately need myth and dogma, and scientists could not explain why people have this need, [Page 61]. To call religion instinctive means that its sources are deeper than ordinary habit and are in fact hereditary, urged into birth through biases in mental development encoded in the genes, [Page 257]. The idea of a genetic and evolutionary origin of moral and religious beliefs will be tested by the continuance of biological studies of complex human behavior, [Page 264]. Human social existence, unlike animal sociality, is based on the genetic-propensity to form long-term contracts that evolve by culture into moral percepts and law. How did this genetic propensity originate? Either it was given to humanity from above, or it has emerged randomly in the mechanics of the brain. Still it could have evolved over tens or hundreds of millennia because it conferred survival values upon the genes prescribing them [Page 297].

In his book titled *"Human Instinct"*, Robert Winston concludes a chapter on morality and spirituality by writing, on page 309: "so there may well be a genetic tendency towards religiosity, more pronounced in some people than others, which has long been part of the human condition." [Winston, 2002]

INCEST AVOIDANCE

Incest is having sexual relations with first-kin members of one's relatives, such as brothers and sisters, parents, grandparents, children, grandchildren, aunts and uncles. The destructive consequence of incest is a general phenomenon not just in humans but also in plants and animals. In the plant kingdom scientists have discovered many ingenious tricks, by which the different flowers prevent the pollen of their stamens from pollinating their own stigma. The naked moles are a type of rodents which live in underground tunnels eating plant roots and tubers. It was observed that the females and males of every newborn batch always avoid mating with each other. They do that by a certain scent which characterizes every newborn batch. In all social nonhuman primate species such as baboons and chimpanzees it was observed that both adult males and females will sexually spurn individuals with whom they were closely associated in early life. Primate mothers and sons almost never copulate, and brothers and sisters kept together in their infancy mate much less frequently than do more distant related individuals.

In humans, it is a known general rule that familiarity in childhood breeds sexual contempt when adolescence is reached, particularly as far as brothers and sisters are concerned. There is simply no sexual interest in each other. This elemental response was extensively studied by the Finnish anthropologist Edward A. Westermarck, and first reported in his 1891 masterwork *"The History of Human Marriage"* and the phenomenon came to be called 'The Westermarck-effect'.

The incest avoidance phenomenon was further studied in societies which created social rather than biological brothers and sisters. In Taiwan a long-term study was performed on marriages arranged between sons of rich families and adopted daughters who were raised with the sons in order to become married to each other at adolescence. Although the boys and girls are not genetically related, and in spite of the approval of the society, the Westermarck-effect was quite clear. The children had to be coerced in order to consummate the marriage. Most of these marriages produced much less offspring than usual, and many of them ended in early divorce.

The Westermarck-effect was also proved correct in an unintended experiment. In the early Jewish settlements in Israel, which were called the kibbutz this trait of incest avoidance was also observed. Children from different families were raised in close brother-sisterly fashion. On reaching adulthood marriages were quite rare between children of each settlement.

It is evident that the human brain is programmed to follow a simple rule of thumb: Have no sexual interest in those whom you intimately knew during infancy and early childhood. Few exceptions are recorded in history, mainly in archaic royal families such as Egyptian Pharaohs, who believed that their royal and divine blood should not be allowed to mix with common blood.

In human societies there is a further factor in the subject of incest avoidance, and that is the various social taboos which prohibit incestuous sex. Sigmund Freud claimed that societies invented those taboos in order to prevent the harmful effects on the offspring. His theory postulates that incest avoidance is purely cultural, and that there are no biological, in-born factors at all that prevent incest. If the Westermarck-effect really existed, then no taboos would have been required. It is not easy to see why any human instinct should require legal reinforcement. Orthodox social theory holds that morality is largely a convention of obligation and duty, constructed from mode and custom.

Scientific evidence now leans strongly to the Westermarck effect as the original biological factor involved in incest avoidance. Social regulation has just furthered it by adding taboos, derived from the human rational observation of its harmful effects. The case of incest avoidance clearly illustrates that biologically based moral codes have long preceded any cultural, social, or religious rules.

THE WEIRD BY-PRODUCT

It is not that we just hope that we immortally live somehow after death, neither that we fear later reprisals for our bad deeds. We are simply *told so* by our genes. This genetic hunch or notion is universal. It is present in every culture and in every corner of Planet Earth. An article in TIME magazine, titled *"The God Gene",* October 25, 2004, page 68, says:

> "God is a concept that appears in human cultures all over the globe, regardless of how geographically isolated they are. When tribes living in remote areas come up with a concept of God as readily as nations living shoulder to shoulder, it's a fairly strong indication that **the idea is preloaded in the genome rather than picked up on the fly."** (Emphasis added).

Atheistic reasoning dismisses this hunch or notion, arguing that it is just a weird incidental artifact or a by-product of the evolution of our physically based brain, and that it is irrelevant, and has no significance whatsoever.

I am not advocating any sort of after-life from the wide array represented in various Religions and Creeds, for it could also be something completely different. You can, literally or symbolically, pick whatever suits your inner belief. The scores of sacred scriptures could be true divine revelations, as in Monotheistic Religions. They could be divinely stimulated 'intuitions', or could be personal inspirational wisdoms, as in Buddhism. None of all this makes any difference. I can safely state that: *"MANKIND IS THE RELIGIOUS ANIMAL ON PLANET EARTH."*

The hieroglyphics of Ancient Egyptians describe their perception of how the after-life courts would look. Their drawings show the deceased person standing in front of various judges trying to defend his deeds. All of this took place long before any of the contemporary religions appeared. Judging from ancient temples and monuments, I believe that similar ideas existed in all ancient civilizations. Examples abound in the Indian valleys of the Ganges and Indus rivers, the valleys around the Yellow river in China, and the temples in the mountains of South America. In Spain, France, Britain, and Sweden there are ancient megalithic monuments, remains of an ancient civilization that built large stone temples for Gods whom we don't now know. As it happens, the only ancient language that we could decipher is the Egyptian Hieroglyphic.

THE DUALISM BELIEF

As Yale University psychologist Paul Bloom argues in his intriguing book, *"Descartes' Baby"*:

> "We are natural-born dualists. In other words: we are genetically hard-wired to believe in the 'after-life of a Soul that outlives a deceased body. The Intuitive belief in after-life is the origin of all religious feelings." [Bloom, 2004]

In the Sept. 2004 issue of *"Scientific American"* magazine, page 24, Michael Shermer says that the dualist belief that body and soul are separate entities is natural, intuitive and with us from infancy. He gives his personal explanation of this by saying:

> "The reason dualism is intuitive is that the brain does not perceive itself, and so wrongly ascribes its consciousness and subjective feelings to another separate source."

Earlier, the concepts of free-will, and religious feelings were discussed. Every human being feels deep in his heart that he will somehow account for what he has said or done, throughout his life. Same as other instinctive traits, this trait also varies in degree from one man to another; but it is always there.

These facts are instinctive, and are totally irrespective of any social nurture. It is purely personal, and has nothing to do with what *others* feel. It is as private as other subjective feelings such as love, hate, envy, jealousy, pain, anger, and happiness. This mystic feeling cannot be explained by logic or scientific words, but it is part of our genetic make-up, the same as inventiveness and the same as language syntax.

In the instinct of language, the only acquired part is the vocabulary of the society one was born into. Same also with this instinct, the only acquired part is the religion, creed, or moral code of the society into which one happens to be born.

This is the basis of religion, without having to go into any detail. This religious human trait is very similar to the other traits, such as private-ownership and aggression. All of them can be expressed in either good or evil ways. They can be most destructive if their ways of expression are only channeled through evil means.

The 'acquired' part of the language instinct becomes part of the self-image of a tribe. It can cause wars. One glaring example is the war which divided Pakistan into east, (Bangladesh), and west. In spite of having the same official religion, the main cause of the conflict was the unwise decision by Mr. Jinnah to make his Urdu language the only official one, completely disregarding the Bengali language of the east. Similarly, the acquired part of our religious instinct becomes a part of the self-image of the tribe, group, or Nation. It can, and did, and does, and will, lead to wars. If the instinctive ness of both these factors is understood, and respected, two major sources of wars may be suppressed and prevented, in a few generations time.

Karen Armstrong wrote many books about the history of almost all religions. She came to the conclusion that Human Beings are "*spiritual*" animals. In her book [*A History of God*], page 3, she even suggested another adjective for Homo Sapiens: and that is "Homo Religiosus"; [Armstrong, 1993].

In our eternal quest for peace we have to accept that religion is here to stay. This is in complete contrast to the vain and hollow crusade of the atheist camp. It is one of many basics of human nature that cannot be eradicated. All we should do is define means for its expression in ways other than feuds or wars.

> > > > >

RECENT SCIENTIFIC RESEARCH INTO BRAIN ACTIVITY, CLOSELY RELATED TO THE SUBJECT OF GENETIC RELIGIOSITY

Probing the mysteries of the human mind has recently become a most fascinating frontier of science. Researchers in this field are haunted by two main questions. First, whether our consciousness and memory could be explained on purely physical grounds? The second question is how to explain human subjective and spiritual feelings if our consciousness is purely physical?

One representative book on these subjects is titled: *"WHY GOD WON'T GO AWAY?,* [Newberg, 2001]. Another book is titled *"PHANTOMS in the BRAIN",* [Ramachandran, 1998]. The subject is quite hot nowadays. It has already reached prestigious lay-people magazines such as Newsweek, *"Searching for the God Within",* an article in the issue of Feb. 5 / 2001, and also Reader's Digest, *"Searching for the Divine",* an article in the issue of December 2001; just to mention a few.

From the many books I read, I have chosen relevant points from the above two books because the writers are respected authorities in their fields. Andrew Newberg is a Professor of Radiology at the University of Pennsylvania. His co-author Eugene D'Aquili is a Professor of Psychiatry at the same University. V. Ramachandran is a renowned Professor of Neurobiology at the University of California.

Chapter 9 of *"Phantoms in the Brain"* is titled "God and the Limbic System". It is full of most interesting points:

The human belief in the supernatural is so widespread all over the world that it is tempting to ask whether the propensity for such beliefs might have a biological basis. Many traits, (e.g. inventiveness and language), make us uniquely human, but none is more enigmatic than religion -- our propensity to believe in God or in some higher power that transcends mere appearances. It seems unlikely that any creature other than humans can ponder the infinite, or wonder about "The Meaning of It All". Atheists, (and communists), say that this is only wishful thinking, and a longing for immortality. "The sick and poor just seek solace in religion!"

Albert Einstein repeatedly described a "Sense of Religious Feeling which shows no Dogma". It overwhelmed him whenever he got drowned in his research into the cosmos and science. *[N.B.: In the next page I'll describe the effects of "meditation", which requires the totally abandoned concentration on one single object until there is a feeling of "unity with the universe". Einstein's revelations surely fall under this heading].*

Every medical student is taught that patients with epileptic seizures originating in the left temporal lobe can have intense spiritual experiences during the seizures. They also become preoccupied with religious and moral issues, and a "sense of mission", even during the seizure-free periods. It is worthwhile to note that epilepsy has affected many great thinkers, sages, and leading figures in the history of Mankind.

Prof. Ramachandran also discusses meditation. He says that Normal people get occasional spiritual **'highs'** and occasional glimpses of what they call **'A Deeper Truth'**. This often takes place while listening to some especially moving passage of music, or while admiring a beautiful piece of art or natural beauty. Meditation requires total concentration on one single object or thought. Experienced meditationists often describe reaching heights of spiritual and religious ecstasy, comparable to what is described by mystics and the Sufis of various religions. Simultaneous stimulation of the smell center, using incense for example, greatly enhances the spiritual state. It does not block the total concentration needed to reach it. It is worthwhile to note that incense and similar perfumes almost always exist in the Houses-of-worship of all religions.

A modern tool for scientific research is called "Trans-cranial Magnetic Stimulator". When applied to the scalp it shoots a rapidly fluctuating and extremely powerful magnetic field that can be focused on any small patch of brain tissue thereby activating it, and providing hints about its function. This is done with the patient fully conscious, and so he can easily describe any subjective feelings. If we stimulate a motor area, the related muscles start twitching. If we stimulate a sensory area, the patient describes what he feels. This technique, as well as 'positron emission tomography' and 'functional magnetic resonance imaging', have been used to study the brain changes taking place in expert meditationists before, and when, they reach their "highs".

The Limbic System is located on the undersurface of the forebrain. It comprises the structures called the hippocampus, amygdala, septum, thalamic nuclei, mamillary bodies, and cingulated cortex. Our sense of smell is centered in this area; other known functions are emotional associations of events and memories. When these areas get stimulated by epilepsy, *or artificially*, the most striking symptoms are emotional. Women sometimes experience "an overwhelming orgasm of body and soul". But most remarkable of all are those who describe deeply moving spiritual experiences, including a feeling of divine presence, and the sense that they are in direct communion with God. "Suddenly it all makes sense, I feel **ONE** with the whole cosmos", is a common expression, Prof Ramachandran writes.

> > > > >

Dr. Newberg decided that religious experience is too intriguing not to study. He has given the name of 'Neuro-theology', or 'The biological Theory of Religion', to the new research discipline. The following are chosen points from his book *"WHY GOD WON'T GO AWAY?"* :

The human brain has been genetically wired to encourage religious beliefs. In his research, an inescapable conclusion is that GOD IS HARDWIRED INTO THE HUMAN BRAIN.

In repeated experiments, using an imaging technology called 'SPECT', Dr. Newberg documented visible organic change in the blood flow of a chunk of the brain's left parietal lobe called the orientation-association area. This region is responsible for drawing the line between the physical self and the rest of existence, a task that requires a constant stream of neural information from the five senses. This sensory flow is diminished, or abolished, during deep meditation. The same also takes place by the total concentration on enjoyable experiences such as a particularly moving piece of music, or fabulous scenery of natural beauty. That's why even non-believers are often moved by religious rituals and hymns. At those moments there is an almost total blackout of the orientation area, coinciding with intense spiritual peaks, and a sense of a limitless awareness into infinite space.

The research shows that those feelings are rooted not in emotion or wishful thinking, but in the genetic wiring of the brain. It may mean that God is just a perception generated by the brain, *OR* that the brain has been wired to experience the reality of God, writes Dr. Newberg.

Here I may pose a very intriguing question: If our brain has been wired to experience the reality of God; then surely it is just natural; because then God must have been *the-master-electrician* of the neurons of the human brain, isn't it ?!. This research has still a long way to go, but it is exciting that one can even begin to address questions about God and spirituality in a scientific way.

The knowledge that 'Faith' is somehow hardwired into the human genome gives to Mankind a third option. At present they have no choice except two: Atheism, and Dogma. The bleak spiritual vacuum of atheism is very harmful to the psyche and body of any human. A staunch atheist approaching senility is a very miserable human soul. On the other hand, hundreds of contemporary Scriptures, many of which contradict each other, have ignited many human feuds and wars. This is in spite of the fact that they are all different expressions of the same genetic fact; exactly in the same way of hundreds of contemporary languages being different expressions of our genetic language syntax.

In this book we have achieved 3 important scientific facts:

We made a clear distinction between DNA as an organic material, and GENES as an encoded cipher recorded on the DNA chains. Secondly, we realized that this cipher co-existed with the first appearance of single-celled life on Earth. It has been the mechanism by which the evolution of life proceeded. In this respect genes are similar to the physical laws validated at the moment of the Big Bang, which were the mechanism by which the universe was formed. The evident conclusion is that all of this was a pre-determined master-plan, rather than a random happenstance.

Thirdly, in Chapter 11 we discovered that religiosity, as a part of the human mind, is a built-in code in the human genome.

Dear reader, on scientific basis this is "how" you and I came to be, but we may never ever know "why?"..!

(Figure 20) Human Nature and the World!

REFERENCES

* * * * * * * * *

Aczel, A., **Probability 1: Why There Must Be Life in the Universe.** New York: Harcourt Brace, 1998.

Aitcheson, J., **The Seeds of Speech: Language Origin and Evolution.** New York: Cambridge University Press, 1996.

Armstrong, K., **A History of God**. London: Mandarin, 1993.

Attenborough, D., **The Trials of Life**. London: Collins-BBC Books, 1992.

Barrow, J., **The Artful Universe**. New York: Oxford University Press, 1995.

Benyus, J. M., **Biomimicry: Innovation Inspired by Nature.** New York: Morrow, 1997.

Bloom, P., **Descartes' Baby**. New York: Basic Books, 2004.

Boorstin, D.J., **The Seekers**. New York: Random House, 1998.

Bowker, J., **Is God a Virus?!** London: S.P.C.K., 1995.

Brash, D., and Lipton, J., **The Myth of Monogamy**. New York: Freeman and Co., 2000.

Calvin, W., **How Brains Think.** New York: Harper Collins, 1996.

Chauvet, J.M., **Dawn of Art**. London: Thames and Hudson ltd., 1996.

Colum, P., **Myths of the World**. London: Floris Books, 2002.

Coveney, P. and Highfield, R., **The Arrow of Time**. London: Harper Collins, 1991.

Crick, F., **The Astonishing Hypothesis**. U.K.: Simon & Schuster, 1994.

Damasio, A., **Descartes' Error**. London: Papermac, 1994.

Davies, P., **The Mind of God**. U.K.: Penguin, 1993.

Davies, P., **The Fifth Miracle**. New York: Simon & Schuster, 1999.

Dawkins, R., **The Selfish Gene**. U.K.: Oxford University Press, 1989.

Dawkins, R., **The Blind Watchmaker**. U.K.: Penguin, 1991.

Dawkins, R., **Unweaving the Rainbow**. U.K.: Penguin, 1999.

Dennett, D., **Darwin's Dangerous Idea**. New York: Simon & Schuster, 1995.

Dennett, D., **Consciousness Explained**. U.K.: Penguin, 1991.

Desmond, A. and Moore, J., **Darwin: A Biography**. London: Penguin, 1991.

De Duve, C., **Vital Dust: Life as a Cosmic Imperative**. New York: Basic Books, 1995.

De Waal, F., **Good Natured**. USA: Harvard University Press, 1996.

Doniger, W. (Editor), **Merriam-Webster's Encyclopedia of World Religions**. USA: Merriam-Webster, 1999.

Einstein, A., **The Theory of Relativity.** *"Scientific American"* Magazine, (the second of only two articles on science which Einstein ever wrote for the general public). April 1950.

Eldredge, N., **Reinventing Darwin: The Great Evolutionary Debate.** New York: John Wiley, 1995.

Ferris, T., **The Whole Shebang: A State of the Universe Report.** New York: Simon & Schuster, 1997.

Feynman, R.P., **The Meaning of It All**. London: Penguin, 1999.

Golding, W., **Lord of the Flies**. U.K.: Faber Ltd., 1954.

Gould, S.J., **Full House**. New York: Harmony Books, 1996.

Gould, S.J., **The Structure of Evolutionary Theory.** USA: Belknap-Harvard, 2002.

Greenfield, S., **The Private Life of the Brain**. London: Penguin, 2000.

Gribbin, J. and Rees, M., **The Stuff of the Universe**. London: Penguin Books, 1995.

Gribbin, M. and J., **Being Human**. London: Phoenix, 1993.

Hamilton, W., **The Evolution of Social Behavior**. Journal of Theoretical Biology, 7: 1, 1964.

Hancock, G., **Supernatural**. London: Random House, 2005.

Hawking, S., **A Brief History of Time**. New York: Bantam, 1988.

James, P. and Thorpe, N., **Ancient Inventions**. USA: Ballantine Books, 1994.

James, P. and Thorpe, N., **Ancient Mysteries.** New York: Ballantine Books, 1999.

Kamalshila, A.M., **Meditation: The Buddhist Way of Tranquility and Insight**. U.K.: Windhorse Publications, 1995.

Kauffman, S., **At Home in the Universe: The Search for Laws of Complexity.** London: Viking, 1995.

Moir, A. and Jessel, D., **Brain Sex: The Real Difference between Men and Women.** U.K.: Mandarin, 1996.

Morris, D., **The Naked Ape.** London: Corgi Books, 1967.

Morris, D., **The Human Animal.** London: B.B.C. Books, 1994.

Newberg, A., D'Aquili, E., and Raus, V., **Why God Won't Go Away: Brain Science and The Biology of Belief.** New York: Ballantine Books, 2001.

Parker, A., **In the Blink of an Eye: How Vision Kick-started the Big Bang of Evolution.** U.K.: The Free Press, 2004.

Pennock, R.T., **Intelligent Design Creationism and its Critics: Philosophical, Theological, and Scientific Perspectives.** USA: M.I.T. Press, 2001.

Penrose, R., **The Emperor's New Mind.** London: Vintage Press, 1988.

Pinker, S., **The Language Instinct: How the Mind Creates Language.** New York: Morrow, 1994.

Pinker, S., **Words and Rules: The Ingredients of Language.** New York: Basic Books, 1999.

Ramachandran, V. And Blackslee, S., **Phantoms in the Brain.** New York: Morrow & Co., 1998.

Ridley, M., **The Red Queen: Sex and the Evolution of Human Nature.** U.K.: Penguin, 1993.

Ridley, M., **The Origins of Virtue.** New York: Penguin, 1997.

Ridley, M., **GENOME.** London: Fourth-Limited Books, 1999.

Russell, B. **POWER.** London: Unwin Books, 1988.

Ryan, W. And Pitman, W., **Noah's Flood.** New York: Simon & Schuster, 1998.

Scantling, S. and Browder, S., **Ordinary Women, Extraordinary Sex.** New York: Dutton-Penguin Books, 1993.

Scientific American., New York: 415 Madison Avenue, N.Y. 10017.

Shermer, M., **Why People Believe Weird Things?** New York: Freeman, 1997.

Smith, J.Z. And Green, W.S. (Editors), **The Harper Collins Dictionary of Religion.** USA: Harper Collins, 1995.

Stewart, I., **Nature's Numbers.** New York: Harper Collins, 1995.

Stewart, I., **Does God Play Dice?** London: Penguin, 1997.

Tudge, C., **The Engineer in the Garden: Genetics, from the Idea of Heredity to the Creation of Life.** London: Pimlico-Random House, 1993.

Von Daniken, E., **Chariots of the Gods**. London: Souvenir Press, 1969.

Walker, A. and Shipman, P., **The Wisdom of Bones: In Search of Human Origins.** New York: Alfred Knopf, 1996.

Warner, R. And Szubka, T., **The Mind-Body Problem**. U.K.: Blackwell, 1995.

Watson, L., **Dark Nature**. London: Hodder & Stoughton, 1995.

Watson, J. D., **DNA: The Secret of Life.** U. K.: Arrow Books, 2004.

Weiner, J., **The Beak of the Finch**. New York: Alfred Knopf, 1994.

Williams, T., **A History of Invention.** New York: Checkmark Books, 1987.

Wilson, E.O., **Consilience**. New York: Alfred Knopf, 1998.

Wilson, I., **Before The Flood: Dramatic New Evidence That the Biblical Flood Was a Real-Life Event**. London: Orion, 2002.

Winston, R., **Human Instinct: How Our Primeval Impulses Shape Our Modern Lives?** London: Bantam, 2002.

Wright, R., **The Moral Animal: Why We Are the Way We Are?** London: Abacus, 1997.

* * * * * * * * *

* * * * * * * *

INDEX

* * * * * * * * *

A

absolute zero 53
Acacia tree 113
Achilles Heel 188
Adaptation 97, 99
Adenine 42, 127
Aggression 203
agnosticism 28
Algae 135
All-or-none-rule 141
amber 37
Ameba 5
amino acids 18, 39, 99, 100, 146
anaerobic bacteria 41
Anger & frustration 177
animal kingdom 14, 101, 114, 115,
     118, 140, 164, 166, 186, 200
Anoxia 230
Anthropomorphic Universe 68
antibiotics 32, 88
antibodies 32
anxiety 210, 228, 229, 232, 233
Ararat 154
Archaic paintings and carvings 160
Archimedes xxx
Artificial selection 5
Astronaut xi
Astronomy 52
Atheism 80, 81, 83, 235, 245
Atheists 83, 233, 235, 242
Atrocities during Wars 201
Attenborough, David 113, 164, 170

B

Bacteria 5, 135
Bats 102, 169
Beauty 212, 213
Biblical Stories 151

Big Bang 39, 56, 57, 58, 63, 66, 67,
     68, 69, 70, 71, 73, 92, 127, 144,
     223, 245, 249
Bilharziasis 109, 111
Black Hole 57, 65
Black Sea 152, 153, 154, 155, 156
blood-brain-barrier 227
Body Display 185
Boredom 211
bower bird 212
Brain 160, 164, 223, 225, 229, 230,
     242, 248, 249
Butterfly 67, 128, 169

C

Cambrian Explosion of life 143
Camouflage 99, 118, 119
Carnivores 108
cells xxx, 5, 6, 9, 11, 14, 15, 16, 17,
     18, 19, 20, 39, 42, 99, 101, 107,
     127, 129, 135, 136, 137, 139,
     147, 214, 227, 229, 230, 231
cellulose 5, 114, 136, 182
chain 7, 8, 10, 13, 44, 45, 93, 99, 146,
     219, 233
Chaos 66, 67, 82
Chimpanzee 129, 131
Chlorophyll 99
Christianity 80, 85, 86
Christmas Island crabs 130, 132, 133,
     134, 136, 169
Chromosomes 5
Clergy 85, 93, 102, 198
cloning 16, 20, 138
Co-evolution 99, 120, 122
Codons, "RNA triplets" 12
communism 85, 181, 197
Concept of Government 179, 196, 197

Concept of Work & Money 192
Consciousness 180, 225, 231, 248
Consilience 183, 225, 236, 250
Continental drift 38
Creationism 87, 249
Creole languages 160, 161
cytoplasm 6, 10, 12, 18
Cytosine 42, 127

D

Dark Energy 57
Dark Matter 57, 59, 60
Darwin, C. 5, 25, 83, 121, 157, 166,
       259
Deism 92, 93, 115, 144, 145
Depression 232
Descartes 168, 225, 240, 247
Determinism 20, 22, 145
Devil 216, 217, 219, 220
Dictator, dictatorship 177, 197, 198,
       204
DNA 5, 6, 7, 8, 10, 11, 12, 13, 16, 17,
       18, 21, 31, 42, 88, 127, 128,
       129, 134, 138, 139, 141, 145,
       146, 147, 219, 233, 245, 250
Dogma 242, 245
Domestic animals 141
Doppler Effect 54, 56, 104
Dr. Jekyll & Mr. Hyde 107
Dreams 163, 180, 189, 204, 212, 215,
       227
Drugs 228
Dualism 240

E

Earth 35, 36, 37, 38, 39, 41, 42, 43,
       44, 45, 46, 48, 51, 52, 55, 58, 59,
       63, 64, 67, 68, 79, 82, 84, 86, 90,
       92, 93, 99, 101, 108, 114, 115,
       116, 121, 127, 130, 134, 135,
       138, 144, 145, 146, 151, 152,
       154, 157, 163, 166, 167, 170,
       172, 173, 189, 199, 209, 211,
       215, 216, 219, 231, 239, 245

Earthworms 44
Echo-location 102
Einstein, A. 248
Electromagnetism 66
Electrons 52
Elephants 170
Elizabeth Taylor 53, 198
Embryological Evidence 138, 140
Empathy 234, 235
Encyclopedia 80, 100, 107, 108, 120,
       135, 185, 248
Endorphins 229, 230
Envy 181
enzyme 16
Ethnicity 208
Eureka xiii, xxx, xxxi
Evil 91, 199, 215
Evolution 5, 25, 72, 83, 88, 89, 97,
       99, 108, 121, 125, 129, 135,
       142, 167, 168, 215, 217, 247,
       248, 249
Evolution Theory 217

F

Facial expression 210, 211
Faith 77, 79, 85, 94, 144, 236, 245
feminine figure 185, 213
Feminism 190
Finches 29
Flood, Noah's 154
Forgiveness 203
Fossils 37, 164
Fraud 84, 91, 171, 172
Free-Will 20, 199, 219
Freemasonry 235
Fundamentalism 86

G

G-SAT Theory 125, 141
Galapagos Islands 27, 29, 31
Galaxy 53, 62
Gandhi 203, 211
Genes 13, 21, 41, 42, 105, 144, 145,
       220

Genesis 27, 86, 151
Genetic engineering 19, 20
genetic religiosity 236, 242
genome x, xi, 10, 14, 15, 17, 18, 19,
        77, 94, 127, 128, 146, 239, 245
Genomics 15
geology 37, 43
Germ line cells 6, 17, 19
God 13, 28, 40, 66, 68, 74, 77, 80, 81,
        83, 85, 86, 90, 92, 93, 94, 99,
        121, 145, 151, 154, 167, 170,
        181, 192, 198, 203, 206, 214,
        219, 220, 235, 239, 241, 242,
        244, 247, 249
Golding, W. 248
Gossiping 209
Gould, S.J. 248
Grameen Banks 194
Grant, Peter and Rosemary 31
Gravity 51, 60, 65, 66
Guanine 42, 127
Gulliver's Travels 218

H

Happiness 227
Hawkins, S. 65, 68, 69, 93
heavy elements 48, 54, 63
Helium 60, 63
Hemoglobin 99
Herbivores 108
Heredity 250
hibernate 141
History 35, 68, 69, 142, 167, 177,
        180, 181, 188, 204, 207, 238,
        241, 247, 248, 250
home-based-work 192
Homicide 204
Hominid Line 166
Hubble 56, 57, 59, 70
Human nature 173
Hydatid disease 106, 107, 110
Hydrogen 52
Hypocrisy 215

I

ice 37, 43, 79, 152, 153, 154, 161,
        165, 167, 229
ice age 43, 152, 161, 165
immunity system 14, 94, 232
Incest-avoidance 237
insecticides 19, 32
Inventiveness 163, 219, 225
Iron 46, 60, 61
Islam 80, 85

J

Jealousy 188
Judaism 80, 85, 87
Junk-DNA 10, 13, 17, 127, 129, 137
Just-So 83, 86, 100

L

Language 3, 12, 13, 29, 46, 79, 137,
        143, 146, 159, 160, 161, 162,
        163, 164, 166, 167, 168, 183,
        200, 206, 208, 211, 212, 215,
        219, 225, 228, 240, 241, 242,
        245, 247, 249
Lenin 85
Leonardo Da Vinci 37, 90
Love 72, 94, 159, 163, 181, 186, 187,
        188, 191, 194, 195, 196, 207,
        209, 210, 211, 212, 213, 216,
        228, 231, 234, 241

M

Macro-evolution 30
Malaria 109
Malraux, Andre' 219
mammary line 141
Mankind 3, 7, 16, 19, 20, 21, 22, 27,
        31, 33, 41, 45, 46, 56, 68, 79, 87,
        89, 90, 91, 92, 93, 94, 104, 109,
        128, 129, 137, 138, 142, 144,
        149, 152, 154, 157, 159, 160,
        163, 164, 165, 166, 167, 168,
        169, 170, 173, 177, 183, 184,

186, 189, 190, 192, 193, 194,
195, 196, 197, 200, 202, 204,
205, 209, 210, 211, 212, 214,
215, 216, 217, 219, 220, 225,
226, 233, 236, 237, 243, 245
Mass-Psychology 200
Meiosis 5, 6
Menial jobs 194
Micro-evolution 30
Mimicry 118
Mind 180, 223, 225, 247, 249, 250
Mitochondria 5
mitosis 5, 6
Morality 199, 235
moth 31, 32, 120
Mountains 39
multi-potent 20
Music 207, 212
mutation 32, 88, 105, 137

N

Nakedness 184
Narcissism 181
Natural selection 31, 101
Nature Magazine 138
neuro-transmitters 11
Neurons 227
neutrino 52, 53, 56
Neutrons 52
Neutron Star 62, 65
Newberg, A. 249
Newton, I. 51, 91
Nitric oxide 229
Nobel Prize 7, 15, 60, 146, 201
Nostalgia 207
Nuclear force, strong & week 66
nucleotide 10, 12, 17, 18
nucleus 5, 6, 10, 12, 16, 52, 53, 58, 60

O

Oceans 43
ova 6
Oxygen 60, 61, 68, 135, 136
Ozone 43, 136, 137

P

Paleontology 37
Panspermia 40
Parasitism 99, 108
Parasol ants 182, 183
Particle-wave-duality 53
Phantoms 225, 242, 249
Pharmaco-genetics 18
plankton xxx, 44, 45, 132
plant kingdom 16, 101, 113, 115, 237
poetry 193, 214, 216, 219
pollen 37, 101, 115, 116, 128, 213,
237
pollination 99, 114, 116, 120, 122
Polygamy 189
Post-modernists 177
Potassium 136
Pride 194, 195, 196
Privacy 209
private-ownership 241
proteomics 15
protons 42, 52, 53, 56, 57, 58, 60, 61,
65, 66, 67
Pulsar 62

Q

Quackery 236
Quantum & Uncertainty Principle 66

R

Radar 102
radioactive elements 52, 53
Raellians 173
Ramachandran, V. 249
rays 43, 52, 54, 55, 65, 72
red blood corpuscles 110
Religion 77, 85, 94, 233, 236, 244,
249
Revenge 202
Rhyme 214
RNA 11, 12, 13, 18, 42, 147
Ruminating animals 114
Russell, B. 181, 201, 219

S

salt 154, 169, 170
Scientific American Magazine 225
Scripture 82, 86
Selfishness 180
Selfish DNA 41
Sex 6, 15, 20, 37, 80, 82, 86, 99, 101, 102, 115, 118, 127, 167, 186, 187, 188, 191, 193, 195, 199, 204, 209, 217, 229, 236, 239, 249
sexually producing 190
Shakespeare, W. 219
Shroud of Turin 70, 84
Sibling-rivalry 205, 207
Snakes 120, 141, 212
Social Insects 128
solar system 38, 43, 64, 127
Somatic cells 5
Somatic genetic therapy 17
Sonar 102, 104
Soul 168, 180, 223, 225, 226, 231, 240
space aliens ix, 173
Species 28, 105, 129
sperm 6, 14, 101
spermatozoa 6
Stars 49
Stem cells 20
stress 69, 229, 232, 233, 234
String Theory 66
sub-atomic particles 66, 91
Supernova 54, 61, 62, 63, 64, 65
survival value 142, 144, 179, 189, 210, 215, 216, 218
Symbiosis 99
Sympathy for the helpless 211

T

tectonic plates 155
teenage-rebellion 205
Telomere 16
Thymine 42, 127
Tool-usage 182
Toti-potent 14, 16, 19
Tribalism 206

U

ultrasonic 102, 104, 105, 169
Universe 38, 48, 52, 54, 55, 56, 57, 58, 59, 60, 62, 63, 64, 68, 69, 70, 71, 72, 73, 90, 91, 92, 100, 121, 129, 131, 144, 163, 169, 247, 248, 249
Uracil 12, 127

V

Vanity 184
Virtue 190, 225, 249
Viruses 42
Voyeurism 209

W

"Windows" for learning 228
Wars 188, 201, 215
Water 38, 43
Watson, J.D. 6, 21, 146, 157, 233, 250
Wedding Celebration 179, 189
Whale 44
Wilson E.O. 223, 225, 237, 250
WWW 46, 231

Z

zoo xxix, xxx, 44, 132

"Look at man's pink skin and puny frame -- how could selection evolve greater weakness without the counter-balancing gift of reason being bestowed first? Man must have had the human proportions of mind before he could afford to lose the bestial proportions of body."

Charles Darwin

www.ingramcontent.com/pod-product-compliance
Lightning Source LLC
Chambersburg PA
CBHW031825170526
45157CB00001B/193